AF757428

THE MEXICAN WAR DIARY AND CORRESPONDENCE OF GEORGE B. McCLELLAN

The Storming of Chapultepec, Sept. 13th. 1847,
ca. 1848, lithographer unknown. *Library of Congress.*

THE MEXICAN WAR DIARY AND CORRESPONDENCE OF GEORGE B. McCLELLAN

EDITED BY
THOMAS W. CUTRER

Louisiana State University Press Baton Rouge

Published by Louisiana State University Press
Copyright © 2009 by Louisiana State University Press
All rights reserved
Manufactured in the United States of America
FIRST PRINTING

DESIGNER: Barbara Neely Bourgoyne
TYPEFACES: Minion Pro, text; Mailart Rubberstamp, display
PRINTER AND BINDER: Thomson-Shore, Inc.

The letter from George B. McClellan to Charles Stewart, 1 December 1846, in the Charles Stewart Papers, bMS Am 1243 (509), is reproduced by permission of the Houghton Library, Harvard University. Portions of the letter from George B. McClellan to Samuel Barlow, 8 November 1861, BW Box 37 (32), are reproduced by permission of The Huntington Library, San Marino, California.

LIBRARY OF CONGRESS CATALOGING-IN-PUBLICATION DATA

McClellan, George Brinton, 1826–1885.
The Mexican War diary and correspondence of George B. McClellan / edited by Thomas W. Cutrer.
p. cm.
Includes bibliographical references and index.
ISBN 978-0-8071-3451-1 (cloth : alk. paper) 1. Mexican War, 1846–1848—Personal narratives, American. 2. McClellan, George Brinton, 1826–1885—Diaries. 3. McClellan, George Brinton, 1826–1885—Correspondence. 4. Soldiers—United States—Diaries. 5. Soldiers—United States—Correspondence. 6. Military engineers—United States—Biography. 7. Military engineering—United States—History—19th century. 8. Mexican War, 1846–1848—Engineering and construction. 9. United States. Army. Corps of Engineers—History—19th century. 10. Mexican War, 1846–1848—Regimental histories—United States. I. Cutrer, Thomas W. II. Title.
E411.M127 2009
973.6′2092—dc22
[B]

2009007674

The paper in this book meets the guidelines for permanence and durability of the Committee on Production Guidelines for Book Longevity of the Council on Library Resources. ♾

CONTENTS

THE
MEXICAN WAR DIARY
AND CORRESPONDENCE OF
GEORGE B. McCLELLAN

INTRODUCTION

On 11 May 1846, Congressman William Sawyer, an ardent Jacksonian Democrat (and a former blacksmith) took the floor of the House of Representatives to denounce the "wasteful squandering of the public money upon a pensioned and useless institution," the U.S. Military Academy. Railing against "a longer continuance of these heavy annual appropriations for the support of a few rich men's sons at West Point," he demanded the immediate closing of the school.

At this dramatic moment James K. Polk's messenger entered the House chamber, bearing the president's war message to Congress. "The cup of forbearance had been exhausted," Polk declared. Mexico, he alleged, "has passed the boundary of the United States, has invaded our territory and shed American blood upon the American soil." With hostilities with Mexico now underway, "notwithstanding all our efforts to avoid it," the president called upon the Congress "by every consideration of duty and patriotism to vindicate with decision the honor, the rights, and the interests of our country" by an immediate declaration of war.

Following this electrifying announcement, Sawyer continued with a vengeance. West Point, he declaimed, was "a blot upon our public history," and he predicted that the "rotten and corrupt establishment" on the Hudson would soon "cease to drain the public treasury." Not one quarter of the cadets "who share the smiles and favors of the public school, sustained, as it is, by the people of this country—by the honest, hardy, industrious sons of liberty," he maintained, ever entered military service.

Of those graduated cadets who had entered service, many, "at the trying moment when danger threatened" in the war against the Seminole Indians, had resigned their commissions rather than face a campaign in the Florida swamps. "We want no such truckling pensioners in the contest now raging upon the Rio Bravo," Sawyer thundered. "Hardier sons, and bolder spirits, and more brawny arms, will be there."

Moreover, and more damning still, the congressman claimed that in America's wars, "not a solitary instance is now within my recollection of a brave and gallant champion upon the tented field of slaughter, when desolation and death beat in stormy fury around our hearths and our homes, who ever hailed from this farfamed and fostered institution." Rather, he maintained, whenever "the standard of liberty, shattered and torn by the storm of battle, has been borne in triumph from the field, over the bodies of the slain, it has been cherished by the hand of him who rushed from the plough handle or the work shops of the country to its protection and defence."

As evidence Sawyer pointed to the hero of the hour, Brig. Gen. Zachary Taylor, "Old Rough and Ready," the victor of Palo Alto and Resaca de la Palma, who "came up self-made and self-taught, amid the hardships and privations of a frontier life, to the rank and reputation he now enjoys."

Although his motion to disestablish the academy went down in defeat, Sawyer yet believed that the debate should serve as a warning to "the favored young spirits now there, that the period may not be far distant when those of us who regard it as an incubus upon the country may finally triumph."[1]

The United States' war with Mexico laid forever to rest the fallacy that the graduates of West Point were mere "truckling pensioners" and products of a "rotten and corrupt establishment." To the rigorous education that his engineers received at West Point, Winfield Scott, general in chief of the U.S. Army and commander of the force that was to capture Mexico City, attributed his army's success. "I give it as my fixed opinion that but for our graduated cadets the war between the United States and Mexico might, and probably would, have lasted some four or five years, with, in its first half, more defeats than victories falling to our share, whereas in two campaigns we conquered a great country and a peace without the loss of a single battle or skirmish."[2]

Among "the favored young spirits" at West Point on that May morning, and among the foremost of the junior officers who won their spurs in Mexico, was George Brinton McClellan. McClellan was destined to lead one of the most distinguished public lives of his era, first as chief engineer of the Illinois Central Railroad and as president of the Ohio and Mississippi Railroad; then as commander of the Army of the Potomac and general in chief of the U.S. Army; and finally as governor of New Jersey and as the Democratic candidate for president of the United States in 1864, winning 45 percent of the popular vote against Abraham Lincoln.

Already in 1846–48, however, one may clearly see in the newly commissioned second lieutenant the talent, ambition, and arrogance that characterized the engineer, businessman, soldier, and politician that he was to become. McClellan's Mexican War diary and letters show a young man who was courageous, indefatigable, superbly intelligent, and able to draw to him close personal loyalties. They also reveal a character that was contemptuous of those of lesser talents and social class than his own, quick to see conspiracies where none existed, and prone to place upon others the blame for his own shortcomings and to take credit for actions performed by others, if performed at all.

In these letters one also sees the outrageous social snobbery that later manifested itself in his reference to the baseborn Abraham Lincoln as "the original gorilla" and in his breathtaking snub of his commander in chief, when on 13 November 1861, according to presidential secretary and biographer John Hay, McClellan sent his porter to inform the president and the secretary of war that he was not receiving guests when they called at his home.

His racist contempt for Mexicans is appalling. "They are content," he wrote in his diary, "to roll in the mud, eat their horrible beef and tortillas and dance all night at their fandangos." At various points in his letters, he referred to them as "a very dirty, ugly looking set," "a most deceitful and rascally set of devils," "a precious set of cheats," and "the laziest people in existence." The homes of the peasants he considered to be "the most uncomfortable, the dirtiest, most Mexican things you can imagine. There they crowd together men, women, children, pigs, goats, and everything else. No modesty, no decency, and no cleanliness." In his dealings with local farmers, the young officer found, "The only way [to strike a deal] is to bully them a little," and he quite callously offered to bring home to his physician brother "a Mex skull" from the battlefield of Palo Alto. "There are plenty of them there," he wrote with a studied nonchalance, "but it is very hard to carry anything."

McClellan was also "perfectly disgusted" to find himself outranked by volunteer officers, "a soldier of yesterday, a miserable thing with buttons on it, that knows nothing whatever," in the new lieutenant's estimation. His well-known ambition and egotism might be best illustrated by a diary entry, written along the Rio Grande in his first weeks with the army. "I came down here with high hopes," he wrote, "with pleasing anticipations of distinction, of being in hard fought battles and acquiring a name and reputation as a stepping stone to a still greater eminence in some future and greater war." He found not only distinction and promotion but also bitter frustration with superiors who did not measure up to his standard of excellence. He discovered that he was in an army dominated not by the professionally trained military elite of the nation but by the "hardier sons and bolder spirits" who "came up self-made and self-taught" of whom Congressman Sawyer spoke. And he was introduced to the methodical plodding of a military establishment too slow for his own mercurial temperament.

George B. McClellan was born in Philadelphia, Pennsylvania, on 3 December 1826, the son of Dr. George and Elizabeth Steinmetz (Brinton) McClellan. After attending the University of Pennsylvania, he entered the U.S. Military Academy at West Point on 1 July 1842 and was graduated, at the age of twenty, second in the class in 1846—a class that included no fewer than seventeen cadets who were to become general-grade officers in the Union and Confederate armies during the Civil War. Jesse Lee Reno, Darius Nash Couch, Samuel Davis Sturgis, George Stoneman Jr., Dabney Herndon Maury, David Rumph Jones, Cadmus Marcellus Wilcox, Samuel Bell Maxey, and the redoubtable Thomas Jonathan Jackson were among them. George Edward Pickett was the class "goat," graduating in fifty-ninth place.[3]

In his senior year at West Point, McClellan had declared himself "passionately fond of Military Engineering." Second Lt. Gustavus Woodson Smith, a member of the engineering faculty who had taken a personal interest in the talented cadet and whom McClellan considered to be "the kindest friend I have had on the Point," encouraged him to seek service in the engineers. Only the top two or three graduates of any year's class were considered for posting to that *corps d'elite*, and as McClellan wrote to his mother, "everything is expected of them in time of trouble."[4]

Indeed the curriculum at the West Point of the 1840s was dominated by engineering and engineering-related subjects. Cadets took classes in history, geography, ethics, military science, law, French, and Spanish, but mathematics, drawing, and military and civil engineering were the core of their curriculum. As Chief Engineer Joseph G. Totten admitted, although the cadets' education in engineering was excellent, "so great a portion of the time is unavoidably consumed therein, that other branches, of less importance, but nevertheless necessary to a military education"—English, French, history, geography, ethics, rhetoric, logic, and international, constitutional, and martial law among them—were deemed insufficient and "demand more time than can now be given to them." Cadets graduated well versed in the construction of buildings, roads, bridges, canals, and railroads, and the result of this heavy emphasis on engineering was profound. As John Coldwell Tidball, class of 1848 and a distinguished artillery officer, observed, "We were taught with every breath we drew at West Point . . . that engineers were a species of gods."[5]

With the outbreak of fighting along the Rio Grande, Cadet McClellan wrote to his sister, Frederica M. English: "Hip! Hip! Hurrah! War at last sure enough! Ain't it glorious! . . . Well it appears that our wishes have at last been gratified and that we shall soon have the intense satisfaction of fighting the crowd—musquitos and Mexicans and c. You have no idea in what a state of excitement we have been here."

The United States' war with Mexico was the army's first experience in a conventional conflict in forty years, and young officers from throughout the service rushed to take advantage of the opportunity for promotion and glory. As Fitzhugh Lee, the nephew of the most successful and enterprising of Scott's engineers, later observed, "A war with a foreign country was highly exciting and new to most of the army and navy officers, so that applications for service in Mexico rapidly rained upon the War Department, and the Secretary of War had no difficulty in sending to Mexico the most capable officers." Scott, therefore, was able to draw to his staff the services of most the army's finest engineers, with Joseph G.

Totten, John Lind Smith, Robert E. Lee, P. G. T. Beauregard, Zealous Bates Tower, Isaac Ingalls Stevens, John Gray Foster, Gustavus Woodson Smith, and George B. McClellan among them.[6]

Remarking on his uncle's career in Mexico, Fitzhugh Lee observed that "engineers are as necessary to an army as sails are to a ship; they locate lines of battle, select positions for the artillery, make reconnoisances, and upon their reports the movements of the army are based. They draw topographical maps, construct roads and bridges, and guide troops in battle to positions they had previously reconnoitred."[7]

Scott took full advantage of the intelligence-gathering skills of his "graduated cadets" throughout his campaign against Mexico City, with one of them, P. G. T. Beauregard, recording that the commanding general "never on a single occasion, gave an order for an attack or an important movement when near the enemy, without first having received the reports of his Engineer officers."[8]

In addition to "the examination of all routes of communication by land or by water," the *General Regulations for the Army, 1841* included among the roles of the engineer "the construction of military roads and permanent bridges connected with them, and . . . of field works, for the defense of encampments, fords, ferries, and bridges." The brunt of this duty was to fall upon Company A, Engineer Troops, the first unit of enlisted sappers and miners in the U.S. Army and the company to which McClellan was attached in Mexico.[9]

The collected letters and diary from the early career of George B. McClellan illustrate the intersection and intertwining of the influence of the professional training that he and his brother regular officers received at West Point, the importance of the U.S. Army Corps of Engineers to the victory in Mexico, and the vital mission of Company A, the unit that McClellan helped raise, train, and lead.

EDITORIAL METHOD

As Stephen Sears observes in his biography of McClellan, the young officer "worked up his Mexican War journal periodically," recalling the activities of a week or more in a single entry as the time or inclination allowed. A copy he made in a leather-bound journal, probably in Mexico

City in the winter of 1847–48, is the version preserved in his papers in the Library of Congress. Thus McClellan's manuscript and a number of mostly satirical drawings is neither exactly a diary nor a memoir but something of a blend of the two.[10]

As Sears has written, "a heavily edited transcription by William Starr Myers was published in 1917 under the title *The Mexican War Diary of George B. McClellan.*" The present effort represents a new transcription of the diary, reintroducing those passages that Starr deleted and correcting numerous mistranscriptions and inaccurate annotations.[11]

Published here for the first time are also McClellan's letters to family members, to fellow officers, and in one case to a political official, from West Point and from Mexico regarding the war and the company of engineers that he helped recruit, organize, and lead. I have made the determination to combine the diary and letters into one chronological whole rather than to present them individually. Where I deemed it expedient, I have broken up those parts of the diary that covered a lengthy period of time into daily entries so as to allow the intersession of McClellan's letters. This practice, I feel, has the advantage of maintaining a continuous flow of narrative, the two components of which are mutually reinforcing, shedding light on each other.

In a few instances I have chosen to elide material from his letters to family members that I have deemed of little or no interest to the most likely readers of this book. In such cases I have indicated the elision with a short row of asterisks (*****).

McClellan's penmanship is remarkably clear, and his spelling is consistently good in general, reflective of his excellent education at the University of Pennsylvania and at West Point. But due to the exigencies of field operations, the haste in which he was often writing, and the lack of suitable writing space, he makes frequent use of abbreviations—"WP" for West Point, "comp." for "company," and "Phila" for "Philadelphia," for example—and the ampersand. He almost invariably uses numerals in place of the written number, his punctuation is irregular (in particular he uses dashes in place of periods), and his paragraphs tend to be consistently run on. He is highly inconsistent in spelling such words as "color," "favor," and "endeavor," sometimes with and sometimes without the "u" favored in traditional British usage, occasionally spelling the same word both ways in a single paragraph. The names of ships he sometimes places in quotation marks, though more often does not. Too, although he was fluent in French, well read in Greek, and at least passable in German, his Spanish was not always flawless, and his spellings of Mexican place names are sometimes irregular.

I have undertaken, therefore, to regularize to some degree the mechanics of his writing and to make them more consistent with modern usage. My belief is that this emendation of McClellan's style in no way alters his meaning or mars the flavor or immediacy of the diary and letters but, I hope, makes them more accessible to the modern reader.

Shortly after the battle at Cerro Gordo, McClellan's theretofore copious, detailed, and colorful diary entries came almost to a full stop. Why he gave up what seems to have been an enjoyable writing exercise cannot now be known, but his activities can be closely followed by his letters to his family and by the many references to him in the letters and diaries of his brother officers and in official reports. I therefore have sought to piece out the final months of his Mexican War career from a variety of sources, in addition to his personal correspondence.

Those annotations for which no source of information is specifically cited were informed by such standard reference works as Francis B. Heitman, comp., *Historical Register and Dictionary of the United States Army, from Its Organization, September 29, 1789, to March 2, 1903,* 2 vols. (Washington, D.C.: U.S. Government Printing Office, 1903); George Washington Cullum, *Biographical Register of the Officers and Graduates of the U.S. Military Academy,* 3d ed., 3 vol. (Boston: Houghton Mifflin, 1891);Guy V. Henry, comp., *Military Record of Civilian Appointments in the United States Army* (New York: D. Van Nostrand, 1873); Fayette Robinson, *An Account of the Organization of the Army of the United States* (Philadelphia: E. H. Butler, 1848); Donald S. Frazier, ed., *The United States and Mexico at War* (New York: Macmillan, 1998); Ezra J. Warner's *Generals in Gray* and *Generals in Blue* (Baton Rouge: Louisiana State University Press, 1959 and 1964); Ron C. Tyler, ed., *The New Handbook of Texas* (Austin: Texas State Historical Association, 1996); K. Jack Bauer and Stephen S. Roberts, comp., *Register of Ships of the U.S. Navy, 1775–1990: Major Combatants;* and James L. Mooney, ed., *Dictionary of American Naval Fighting Ships,* 9 vols. (Washington, D.C.: Naval Historical Center,

1959–91). The letters, memoirs, and official reports of major participants in McClellan's campaigns in Mexico—most notably those of P. G. T. Beauregard, Isaac Ingalls Stephens, Dabney Herndon Maury, and G. W. Smith—are listed in the bibliography as are the most useful biographies and scholarly histories of the war with Mexico. Foremost among the latter are Stephen Sears's *George B. McClellan: Young Napoleon* (New York: Ticknor and Fields, 1988), Justin H. Smith's classic *The War with Mexico* (2 vols., New York: Macmillan, 1919), and K. Jack Bauer's *The Mexican War, 1846–1848* (New York: Macmillan, 1974).

1. *Appendix to the Congressional Globe,* 29th Cong., 1st sess., 1848, 585–87.

2. Quoted in Russell F. Weigley, *History of the United States Army* (New York: Macmillan, 1984), 185.

3. John C. Waugh, *The Class of 1846: From West Point to Appomattox: Stonewall Jackson, George McClellan, and Their Brothers* (New York: Warner Books, 1994).

4. Of the 1,365 cadets who by 1848 had been graduated, 84 had been commissioned into the Corps of Engineers and 12 into the topographical engineers. "Report of the Chief Engineer," in *Report of the Secretary of War,* 31st Cong., 1st sess., 1849, H. Exec. Doc. 1, 294.

5. James L. Morrison Jr., *"The Best School in the World": West Point, the Pre-Civil War Years, 1833–1866* (Kent, Ohio: Kent State University Press, 1986), 23–24, 91–101; Totten to William L. Marcy, "Report of the Chief Engineer," in *Report of the Secretary of War,* 29th Cong., 1st sess., 1846, S. Exec. Doc. 1, 134–37; James L. Morrison Jr., ed., "Getting through West Point: The Cadet Memoirs of John C. Tidball, Class of 1848," *Civil War History* 26, no. 4 (Dec. 1980): 323.

6. Fitzhugh Lee, *General Lee: A Biography of Robert E. Lee* (New York: D. Appleton, 1894; reprint, with a new introduction by Gary W. Gallagher, New York: DaCapo, 1994), 33.

7. Ibid.

8. Pierre Gustave Toutant Beauregard, *With Beauregard in Mexico: The Mexican War Reminiscences of P. G. T. Beauregard,* ed. T. Harry Williams (Baton Rouge: Louisiana State University Press, 1956), 46.

9. U.S. War Department, *General Regulations for the Army, 1841,* art. 75, paragraph 878, 159.

10. Stephen Sears, *George B. McClellan: The Young Napoleon* (New York: Ticknor and Fields, 1988), 413n19.

11. Ibid.

1
"WAR AT LAST SURE ENOUGH!"

WEST POINT

18 March 1846–22 September 1846

With Texas statehood, Pres. James K. Polk sent Zachary Taylor to the Nueces River in command of the so-called Army of Observation. When negotiations with Mexico for the sale of California failed, he ordered Taylor to the Rio Grande with the army, now styled the "Army of Occupation," precipitating war with Mexico. Taylor's command defeated Mexican forces under Mariano Arista at the battles of Palo Alto (8 May 1846) and Resaca de la Palma (9 May 1846), both in southern Texas.

On 1 July, seven weeks after the war's first battles, McClellan received his commission as a brevet second lieutenant of engineers and was granted a furlough to be spent in Philadelphia. Having been there but a few days, however, he received orders to return to West Point to take part in the organization and training of a new company of engineering soldiers. Such an assignment was "all that I could hope, ask, or expect," he wrote to Frederica; "it is exactly what I desired."[1]

Since the Army Reduction Act of 1842, the army had no engineer troops, but on 15 May 1846, only days after the first fighting along the Rio Grande, Congress authorized the addition of a company of "sappers, miners, and pontoniniers" to compose a part and to be led by officers of the Corps of Engineers. The unit's functions were to include "all the duties of sappers, miners and pontoniers," and it was also to "aid in giving practical instructions in these branches at the Military Academy." The enlisted organization comprised ten sergeants, ten corporals, two musicians, and seventy-eight privates.[2]

During the first half of the nineteenth century, the enlisted men of the regular army were predominately of foreign birth and recent immigration, and in the estimation of Isaac Ingalls Stevens, who had graduated at the head of the West Point class of 1839 and commissioned in the Corps of Engineers, they were "far inferior in character" to what they should be and what they might become. For the engineer company, therefore, he advocated recruiting only young men of American birth, intelligence, good character, and above-average stature. Like Cromwell's Ironsides, the engineer troops should be enlisted solely from among men of "good family and substance," and Falstaff's "serving-men and tapsters" ought to be scrupulously avoided.[3]

These recruits, he urged the War Department, should undergo a thorough course of instruction in combat-engineering techniques in order "to raise then to the highest state of discipline and efficiency, a fair representation of what an American army might and should be, so that every man in the company can, if he chooses to study and do his duty, become a good clerk, overseer, or practical engineer." Thus, Stevens believed, service in the U.S. Army could be made "as honorable for the common solder as for the officer."[4]

The company was recruited "by great exertions on the part of several engineer officers," Col. Joseph G. Totten, the chief of the Corps of Engineers reported, and was placed under the command of Capt. Alexander Joseph Swift of North Carolina, who had graduated first in the West Point class of 1826 and subsequently detached to attend the French army's engineering school at Metz. As his second in command, Swift selected McClellan's mentor, Gustavus Woodson Smith, and at Smith's urging he selected McClellan as the company's third and junior officer. Smith "felt assured that he would be in full harmony" with Swift and McClellan in the leadership of the engineer troops. "No three officers of a company of soldiers ever worked together with less friction," he later wrote, and "the understanding between them was complete. There were no jars—no doubts or cross purposes—and no conflict of opinion or action."[5]

In his quarters at West Point during the summer of 1846, Swift personally instructed Smith and McClellan in the engineering techniques that he had learned at Metz, and his lieutenants in turn drilled and exercised the men and taught their captain the new manual of infantry tactics that had been introduced to the army since Swift's graduation from West Point. In the words of George Stillman Hillard, McClellan's official biographer during his presidential campaign in 1864, "The duties in which Lieutenant McClellan now found himself engaged were very congenial to him, and he devoted himself to them with characteristic ardor and perseverance."[6]

1. Sears, *George B. McClellan,* 13.

2. Theophilus F. Rodenbough, *Army of the United States: Historical Sketches of Staff and Line with Portraits of Generals-in-Chief* (New York: Maynard, Merrill, 1896), 116.

3. Isaac Ingalls Stevens, *Campaigns of the Rio Grande and Mexico* (New York: D. Appleton, 1851), 13. See also Joseph G. Totten, 2 Dec. 1847, in *Report of the Secretary of War,* 30th Cong., 1st sess., 1847, H. Exec. Doc. 8, 67.

4. Hazard Stevens, *The Life of Isaac Ingalls Stevens,* 2 vols. (Boston: Houghton Mifflin, 1900), 1:93–94.

5. Totten quoted in Hillard, *Life and Campaigns of George B. McClellan,* 20; Gustavus Woodson Smith, *Company "A," Corps of Engineers, U.S.A., 1846–1848, in the Mexican War,* ed. Leonne M. Hudson (Kent, Ohio: Kent State University Press, 2001), 1–2.

6. Hillard, *Life and Campaigns of George B. McClellan,* 14.

George B. McClellan to Elizabeth Brinton McClellan[1]

18 MARCH 1846

UNITED STATES MILITARY ACADEMY

I received a few days ago, my dear Mother, your kind letter and now avail myself of what is really the first opportunity of answering it between plenty of hard work in the line of Engineering, Ethics, Artillery, &c, &c. during the week and being obliged to spend my Saturday evening at the Dialectic, I have but little leisure, you may imagine.[2] Things go on here pretty much in the same old style, but the great blessing is that time flies quickly. I have become pretty well tired of my four years' slavery, and long to be free once more; it really seems almost incredible that I shall graduate, be my own master, and strike out to make my own name and fortune in three months! Nevertheless it is true! I am not afraid to try it since all of you have observed the most profound silence with regard to the questions I asked about the "Observatory" business. I have concluded to make up my mind about the matter at once and have done with it. This with all due deference—I find that it is quite certain that I can enter the Corps of Engineers and have, with the advice of an officer in whom I have the greatest confidence, resolved to apply for that Corps. It holds in every respect the very highest rank in our service; it affords the very best opportunities of making one a thorough Engineer; and as its officers carry to a great extent the study of Military Science, as well as that of Engineering, everything is expected of them in time of trouble. Now, I am passionately fond of Military Engineering and Military studies. I can, in this Corps, indulge this propensity. I can make myself a better officer than I would be entering the Line, and I will be more fit to take up the study of Law, should I at any time resign my commission. The "Engineer" is not a "Staff Corps." It is more essentially a Military Corps than any other in the Army. These are my reasons for applying for the Engineers. Under these circumstances I cannot accept the offer made by Mr. Walker,[3] for the Chief Engineer will not permit his officers to be detailed on any such duty, since the number of officers in the Corps is not sufficient for their duties. And even if he would consent, which is not possible, I must confess that I would not like the nature of the service; I should like to be stationed at home, and that would be my *only* reason for accepting the offer.

I shall *probably* be sent to Texas next fall, but if Congress ever does pass the "Pea Patch" appropriation, might be stationed there, in the course of two or three years.[4] I should like to go to Texas, even if there is no war with Mexico, for I would there learn some of the most important duties of an officer. Understand me—I would not be in Camp, as the officers of the Line are, but would be traveling about, either making military "reconnaissances" of the territory, or surveys for sites of Fortifications, &c. If there is a war, I am bound to go there *anyhow.* The amount of it is, that I have seen just enough of Engineering to desire to know more about it. As to studying law, I can say nothing certain about it; it is quite probable that I shall have a shingle up when 25 or 26, but time will have to settle that matter.

I alluded to an officer who had advised me to enter the Engineers. I meant Lt. G. W. Smith, 2nd Lt. of Engineers, and universally acknowledged to be one of, if not *the,* most talented young officers in the service.[5] He "took a fancy" to me in the section room, one day last fall when he heard our section for Mr. Mahan,[6] sent for me to come and see him and his wife, and since then has been the kindest friend I have had on the Post.

The *tailors* have been up here within the past two or three weeks to take our measures. We had one from Boston and one from New York. I ordered of the Boston man, a suit of *undress* Engineer uniform, I won't get my full dress until I am in there for certain. Of the New York one I ordered the "Citz[en]'s" clothes I wanted. So you see, I am beginning to strike out sure enough; there is nothing like having credit to start with. Fortunately our Army officers have quite a different reputation with the tailors &c from that of English and Naval officers. We always pay them, and they are accordingly perfectly willing to trust the graduates. I was quite moderate in my "undress." I had some thought of applying to stay here during the Encampment, in Artillery, but I am afraid that they would not give me no more than a month's furlough in the fall if I did stay, so I do not think that I will buy it. Tell Arthur, that since he is the literary genius of the younger portion of the family, he must manage to write me a postscript telling me what to bring him from N.Y. He must do the same on Mary's behalf, and if they wish anything which is within reach of a brevet second Lieutenant, they shall have it.[7] Tell John,[8] that if he does not write immediately, I will carry into effect the threat I made when I last wrote you. I will write to him—I reckon that will bring him to terms. Tell him that I want to know all about his office, business, &c. &c. I hope this spring weather which is coming at last will help Father with his Neuralgia. Mr. Bailey,[9] Cap'n Keys,[10] &c. all desire to be remembered to him. I have not heard from Frederica[11] lately. I cannot really write to her during the week. We have had a very cold snowy February; the snow has very nearly disappeared, and I hope that we have seen the last of it. Talking of rats reminds me that Joe Folsom told me that he dined with you when he was in the city. What did you think of him? I suspect he is trying to "ring in" to Miss Tevis.[12] I hope he may succeed. Give my best regards to Aunt Steinmetz, Judge Coxe, &c., &c.[13] The 15th was your birthday, was it not, Mother? I only wish that you may see plenty more of them and happy ones too. I'll do all that is in my power to make them so—rest assured of it.

1. Elizabeth Steinmetz (Brinton) McClellan was born on 15 March 1800, the daughter of John Hill and Sally (Steinmetz) Brinton, a socially prominent Philadelphia family. On 14 September 1820 she married Dr. George McClellan. They were the parents of five children: sons John Hill Brinton, George Brinton, and Arthur, and daughters Frederica and Mary. Elizabeth Brinton McClellan died on 18 August 1876.

2. The Dialectic Society was the only sanctioned extracurricular club at West Point during McClellan's time there and was, as John C. Tidball described it, "a musty organization among the cadets" that represented "the only thing approximating literary exercises" there. McClellan served as the club's president during his senior year. James L. Morrison Jr., ed., "Getting through West Point: The Cadet Memoirs of John C. Tidball, Class of 1848," *Civil War History* 26, no. 4 (Dec. 1980): 320.

3. In 1840 Nears Cook Walker, one of the nation's most important early astronomers, had offered McClellan a position at his Philadelphia observatory, at that time one of the two finest in the United States.

4. Fort Delaware, constructed on Pea Patch Island in the Delaware River in 1847, was designed to protect Philadelphia and its harbor. At the time it was the largest coastal defense installation in the country. McClellan was stationed there from June 1851 until March 1852.

5. Gustavus Woodson Smith, class of 1842, was commissioned into the Corps of Engineers. He returned to the military academy in 1844 as an assistant professor of engineering and on 24 September 1846 was reassigned to the company of sappers, miners, and pontoniers. After a distinguished career in civil engineering, Smith accepted a commission as major general in the Confederate army, becoming second in command of the army that faced his old friend, George B. McClellan, during the Peninsula Campaign of 1862. Leonne M. Hudson, *The Odyssey of a Southerner: The Life and Times of Gustavus Woodson Smith* (Macon, Ga.: Mercer University Press, 1998).

6. Dennis Hart Mahan was graduated at the top of the West Point class of 1824 and was commissioned into the Corps of Engineers. In 1832, after resigning from the army, he returned to the academy, remaining on the faculty for nearly forty years. Mahan redesigned the school's engineering programs, teaching civil and military engineering and compiling several texts long considered standard. Mahan also taught the course on military science taken by virtually every cadet who later fought in the Civil War.

7. Arthur and Mary McClellan were George B. McClellan's younger siblings. Arthur was born on 23 September 1839 in Philadelphia. During the Peninsula and Sharpsburg campaigns, he served as an aide de camp to his brother with the rank of captain. At the end of the war, he was serving as aide to Maj. Gen. Horatio Gouverneur Wright with the rank of major. After the war he went to work for his Brinton family cousins at the Coxe Brothers and Company, coal operators, in Drifton, Pennsylvania.

Of Mary, who was born in 1841, very little is known. Apparently in ill health, she was still living with her mother in 1862. Stephen W. Sears, ed., *The Civil War Papers of George B. McClellan: Selected Correspondence, 1860–1865* (New York: Ticknor and Fields, 1989), 182–83.

8. John Hill Brinton McClellan, the elder brother of George B. McClellan, was born in Philadelphia on 13 August 1823. He received the M.D. degree from the University of

Pennsylvania in 1844, afterward serving as professor of anatomy at Pennsylvania Medical College and in the medical department of Pennsylvania College. During the Civil War he was a surgeon at Philadelphia's South Street Hospital and at Mower's Hospital. Some of the operations that he performed there are described in *The Medical and Surgical History of the War of the Rebellion* (Washington, 1870). John was married to Maria Eldredge in her hometown of Boston on 6 December 1848.

9. Jacob Whitman Bailey, class of 1832, was the professor of chemistry, mineralogy, and geology at his alma mater.

10. Erasmus Darwin Keyes, class of 1832, returned to the academy in 1846 as an instructor. He did not serve in the Mexican War but nevertheless became a favorite of Winfield Scott, becoming the general's aide and military secretary. Keyes commanded the IV Corps, Army of the Potomac, under McClellan during the Peninsula Campaign. He is the author of *Fifty Years Observation of Men and Events, Civil and Military* (New York: Charles Scribner's Sons, 1884).

11. Sarah Frederica (McClellan) English was McClellan's elder sister. She was married to Thomas C. English, a planter from Monroe County, Alabama.

12. Joseph Libbey Folsom, class of 1836, served as an instructor of infantry tactics at the academy. Although he amassed a considerable fortune through land speculation in California, he died a bachelor in 1855.

13. Elizabeth Steinmetz, the maiden aunt of McClellan's mother, was born in 1806 in Northampton County, Pennsylvania, the daughter of John Henrich and Barbara Steinmetz. Her sister, Sally Steinmetz—George B. McClellan's grandmother—married John H. Brinton on 30 April 1795.

Charles Sidney Coxe, the son of Tench Coxe, assistant secretary of the Treasury during Washington's administration, was born in Philadelphia in July 1791 and served as deputy attorney general of Pennsylvania and judge of the District Court for Philadelphia (1826–41). He was also the owner of a chain of Pennsylvania anthracite coal mines for which McClellan's younger brother, Arthur, worked after the Civil War. Judge Coxe married Ann Maria Brinton, the sister of Elizabeth Brinton McClellan.

George B. McClellan to Frederica M. English

UNITED STATES MILITARY ACADEMY

3 MAY 1846

I received a day or two ago, my dear Sister, your last letter, and as I have not been troubled with a remarkably excessive postage account of late, I was right well pleased to get it. I have, *sure enough,* resolved to apply for the Engineers, but if there is any truth in the last accounts from Texas, I shall probably go in the Line, and what's more, be in Texas by the end of next month. The account to which I allude is a rumor, contained in yesterday's papers, to the effect that Gen. Taylor has been defeated with the loss of 700 men—about one third of the Army; if that many have been killed, at least an hundred must have been officers, for great reliance is not placed upon the *present* privates in the Army (they being for the most part composed of these wretched Dutch and Irish immigrants, in time of peace—except the Dragoons who are almost altogether Americans) and as a necessary consequence the officers will have to exert themselves greatly and it is quite certain that no graduate of this Academy will act in a cowardly manner in face of the enemy. If so many officers are killed the whole of our class will be ordered down there at once, to supply the vacancies.[1] If the present state of affairs continues, I shall apply for Texas, whether I get in the Engineers or not, and will I suppose have a chance to pay you a flying visit on the way. I find that the remark I made above about the composition of the Army is a little overdrawn, it is not quite so bad, tho' bad enough. I do not now entertain any doubt as to entering the Engineers, but there is no telling what may turn up with time. You may well imagine what a state of excitement we are in about these times. The Army (!) of Occupation—1900 strong—is in front of Matamoras, which is garrisoned by 7 to 8000 Mexicans, pretty great odds against troops who have never smelled powder in their lives. All of us have friends among the officers, and the first thing we know may be that about half of them are knocked on the head and that we are to go down and enjoy the same amusement! Well! These are among the enjoyments and luxuries of we Army officers, whom we blackguard for a lazy set of goodfornothings!

May 9th. Tempus fugits, with a vengeance, and in the mean time I have been so busy that this letter had to take care of itself. It appears now that the rumor of the fight on the Rio Grande was a humbug after all. The Mexicans have very little idea of fighting, without we commence it; they will be wakened with gasconade. We received news yesterday that Lt. Porter[2] of the 4th was killed by a party of Banditti. I suppose that a good many such things will happen before the affair is settled. Congress has shown their beautiful appreciation of Military matters, by appointing a Committee for the purpose of causing the officers who had those deserters shot when crossing the river tried for *murder*!! If that is the way in which they intend to manage things, they will find it hard work to get officers to remain in the service.

May 13th. Hip! Hip! Harrah! War sure enough! Ain't it glorious! 1500 Regulars and 50,000 Volunteers! Well, it appears that our wishes have at last been gratified and we shall soon have the extreme satisfaction of fighting with the crowd—Musquitoes and Mexicans &c. It now appears that the Government has placed Genl. Taylor in a very dangerous situation, from which, may the Lord deliver him, for it is pretty certain that the Volunteers won't. After all the effervescence of patriotism which the [*word missing*] created at first at New Orleans, it appears that but *650* out of the required 3000 have enlisted—or rather *volunteered*. The amount of it will be that they will have to dispense with the whole volunteer system as far as an Army of invasion is concerned, and trust entirely to the regulars. If 7000 regulars had been at Genl. Taylor's disposal on the Rio Grande, the war would in all probability have been finished at a blow, he would have crushed Arista,[3] and that would have been the end of it. Our class has been expecting orders for the last few days to go to Texas, but I doubt whether we will be sent before we graduate, without the recruiting service goes on so very rapidly that we shall be required to drill, and carry on the recruits to the Army. I am in a very unpleasant state of uncertainty as to my part; I must wait the progress of events for a while, and then decide whether to try it in the Engineers or the Dragoons. One thing is certain, I am determined to go to Mexico if it is possible to work it so. You have seen the account of the gallant exploit performed by Capt. and Lieut. Thornton and Green Mason in cutting their way thro' that detachment of 2000 Mexicans who surrounded them.[4] The only way in which I can account for their losing their men is this—Dragoon officers always charge *literally* at the head of their men—thus the squadron that charges is preceded by its officers who are some yards in advance—now Thornton is one of the best riders and has the best horse in the Army and Green Mason is not much, if at all, behind him, they were probably some distance, say 50 yards, in advance of the men, at the termination of the charge, and when surrounded by the Mexicans, a body of them probably got between these officers and the troops, seeing then that they could not possibly get to the Dragoons, they made a dash at a weak spot and broke through. Tell Tom[5] that it will give me the greatest possible pleasure to accept his offer with helping me to a horse. I shall have to take one—about ¾ths blood about 4½ or 5 years old (wouldn't care if he were only 4 provided he was pretty well advanced for his age) strong, well-made, and active, he's my horse! I cannot tell now when I will be ordered on, but will let you know as soon as I receive my orders, and will certainly stop to see you on the way if possible for me to do so. You may expect to see me sometime between this and October. You have no idea in what a state of excitement we have been here. Everything is topsy-turvy in our heads. I have been and am still very busily employed in writing my *valedictory* as President of the Dialectic Society. The singular appearance of this letter is to be accounted for by the fact that I've concluded to put it in an envelope. Give my love to Tom and tell him not to "list," for volunteering ain't what it's cracked up to be, by a long shot. For we poor regulars who make up our minds to take it as it comes, it goes bad enough, that is, the salt pork, salt water, long marches, wet bivouacs, nights without any sleep, yellow fever fact of it. But the Lord deliver the fancy volunteers. I reckon none of them will try it a second six months.

I have not heard from home in a coon's age. I wonder what will become of "Company H, Heavy Artillery." As soon as I get that valedictory finished, I will write again. Until then—adieu—Your affectionate brother, Geo. B. McClellan

This beats your letter as to looks—all for blunders. How's my namesake? On a paper that I got from home this morning, Mother remarked that Arthur was not fully prepared for the war!

1. This rumor of Taylor's defeat was of course incorrect.

2. Theodoric Henry Porter attended West Point in 1835 and 1836 but did not graduate. He was commissioned a second lieutenant in the Fourth Infantry on 1 July 1839 and was killed in action against Mexican partisans on the Rio Grande, 19 April 1846.

3. Mariano Arista entered the Mexican army in 1819, rising to the rank of general of brigade. At the outbreak of the Mexican War, he was placed in command of the Army of the North but was defeated by the forces of Brig. Gen. Zachary Taylor at the battles of Palo Alto and Resaca de la Palma. He relinquished his command to Francisco Mexía and requested trial by a court-martial, but he was absolved of guilt.

4. Seth Barton Thornton was commissioned into the Second Dragoons directly from civilian life in 1836. He was killed by Mexican artillery fire while on a reconnaissance mission at San Antonio, near Mexico City, 18 August 1847. Winfield Scott noted: "He was a

brave officer and a thorough gentleman, but was always unfortunate in his military career." Quoted in *The Mexican War and Its Heroes . . .*, 2 vols. (Philadelphia: Lippincott, Grambo, 1850), 2:38.

George Thompson "Green" Mason, class of 1842, was commissioned into the Second Dragoons. He was killed in action at La Rosia, Texas, near Fort Brown, 25 April 1846.

5. Thomas Cassander English was born in Clarke County, Alabama, on 28 October 1819. According to his obituary notice, "whilst attending a course of medical lectures at the Philadelphia University, he met, loved, and wedded" Frederica McClellan, "the daughter of the celebrated Dr. McClellan, and the sister of Gen. J. [*sic*] B. McClellan." He was the master of Cedar Hill plantation in Monroe County, Alabama, and by 1860 was the owner of sixty-eight slaves and more than eight hundred acres of land. *The Monroe (Ala.) Journal,* 16 June 1884.

George B. McClellan to Frederica M. English

WEST POINT

16 AUGUST 1846

I received your letter this morning, my dear sister, and very fortunately this is not only Sunday, but also a Sunday for which I am disengaged. I reckon you will hardly believe me, but it is actually the case, that this is the first Sunday since I have been on the Point (since I graduated) that I have to myself. Last Sunday for instance, I was engaged from 10 in the morning until 11 at night (except barely time enough for my meals) in making a drawing of a wagon for our needs. I am actually occupied every day until 10 at night, with drills, company business, &c. &c. for since the men are all raw recruits, and the whole affair is a new thing, both to officers and men, and we have not only to work but also to study, we are also very much pressed for time, as we *shall leave for Mexico early in September.* I am permanently attached to the Company, and as there is a very great probability of their being a battalion of Engineer Troops established, especially if "Company A, Corps of Engineers" distinguishes itself in the ensuing campaign, We have a chance of obtaining higher commissions than those we now hold, if we few distinguish ourselves. I am sorry to tell you that there is every probability (certainty, I might say) of the Company's going by sea, in which case I must accompany it, and cannot pay you the visit to which I have been looking forward with so much pleasure. But the Captain says that he thinks it most probable that we will sail to New Orleans, *thence* either to Matamoras, or Vera Cruz, instead of sailing *direct* to our destination from New York. In this case you might meet me in New Orleans. I do not think this latter method of proceeding very likely to be followed out, but the Captain knows better than I do. I will let you know as soon as we receive positive orders as to our destination. I feel as sorry as you do that I cannot come and see you before joining the Army, but if we do sail for New Orleans, you and Tom must manage it so as to meet me there. Tell Tom—or rather let me tell him, as he reads this—that "I accept with thanks immeasurable the horse you offer. I had intended to take on a horse from the North, but the one at my disposal is only 3½ and could not stand the fatigue of a campaign. Your Leviathan colt would in all probability get on much better since he is used to the climate, and to feeding on grass. I leave it entirely to your good judgment as to whether he will be able to stand it. I shall of course buy a good mustang to ride habitually on the march, and use the horse you give me in *action* and on extraordinary occasions. I will get you to ship him to New Orleans according to a direction which I will send on. Thence, either will take him myself if we sail to that place or a friend of mine will take charge of him for me. Will let you know as soon as I learn, *when* to send him and will take good care not to let him run away with me *in the wrong direction!* Will you let me know as soon as you can his height, size and breadth of chest &c. &c. so that I can get a saddle made to fit him."—The difference since you ask the question, between the Topogs and the Engineers is, that the Engineers are the highest Corps in the Army. Their duties are essentially military, whilst those of the Topogs consists chiefly in making surveys &c. and are of a much lower order.[1] The "Sappers and Miners" are Engineer troops, and as to the General's opinion concerning the "talent" required to officer the "Engineer troops," I must beg leave to differ from him in this, as well as his other opinion (if such really is his opinion). I am attached to the Company of Engineer Troops as 2nd Lt thereof. I'll stay in Mexico as long as the war lasts. My "present situation" is all that I would hope, ask, or expect; it is exactly what I desired—I am the only one of my class (of those who entered the Engineers) who is to go to Mexico, and I must confess that I have malice enough to want to show them that if I did not graduate head of my class,

I can nevertheless do something. I have therefore fully made up my mind either to return from Mexico with at least one more brevet, or else to stay there. I will at all events try not to disgrace our name. "Think on!"

I was very much pleased with the Colonel.[2] He dined at our house one day in company with two of my classmates and we were all delighted with him. Tell that little namesake of a nephew of mine, that if he don't make haste and get well, I will never in the world leave him a single Shilling of my immense fortune. Tell him he will destroy all his brilliant prospects if he don't get along so fast as to talk like a book when I see him.[3] I think that I'll try to ring in to a *Mexican indemnity* before I try the *Alabama fortunes*—what say you?

The flies are almost eating me up alive. I shall have to turn the tables on them presently if they don't keep pretty wide awake. The whole tribe of Miles' (you know who they are) came on here the other day and commenced operations by calling me "George," upon which I took the liberty of calling to their attention the necessity of politeness, and the fact that some people grew familiar on a short acquaintance—*others on none at all,* and the *left* then. I have had two or three such little operations with some of the Doctor's friends, and have made myself *very popular* with them. Do write soon and believe me to be your affectionate brother,
Geo B. McClellan

My respects to the Colonel. *Ask him* whether he has any message to "Old Zack." I'm afraid to write any longer or the flies will eat me up.
Goodbye till next time.
Cadet Geo B. McClellan *is* wrong, it should be *Lieut.* G.B.Mc.

1. In 1816 Congress added a Corps of Topographical Engineers to the army to serve as surveyors, explorers, and cartographers for "the defence of the frontier, inland and Atlantic, and of positions for fortifications." Although in time of war the Corps of Engineers and the Topographical Engineers, or "Topogs" as they were called, engaged in practically identical duties, George Gordon Meade complained that he had been "pretty much a spectator" since his arrival in Mexico, "the Corps of Engineers having performed all the engineering that has been done. This is attributable to the presence of Colonel Totten, who wishes to make as much capital for his own corps, and give us as little as possible." Adrian George Traas, *From the Golden Gate to Mexico City: The U.S. Army Topographical Engineers in the Mexican War, 1846–1848* (Washington, D.C.: Center of Military History, 1993), 177–208; George Gordon

Meade, *The Life and Letters of George Gordon Meade, Major-General United States Army,* 2 vols. (New York: Charles Scribner's Sons, 1913), 1:192–93.

2. This seems to be Lt. Col. John Darrington of South Carolina, commander of the Fourth U.S. Infantry during the War of 1812. In 1854 a John Darrington was a resident of Monroe County, Alabama.

3. During the Civil War, George McClellan English served as a private in E. M. Holloway's independent cavalry company, also known as Crocheron's Light Dragoons, which served as the escort company of the commanders of the Army of Tennessee. He was later a private in the Thirty-first Alabama Infantry.

George B. McClellan to Elizabeth Brinton McClellan

WEST POINT

23 AUGUST 1846

I am afraid that you will think me a sad reprobate, my dear Mother, since I have not written a word to *you,* and very few to anybody else at home, but I assure you that it was not because I did not think about you. I now go to drill before breakfast and to company drill at eight for one and a half hours then to squad drill for another hour and a half, then down to barracks to inspect, and attend to my share of the company business. At 2:30 I go to Engineer drill which lasts until 6:00, then I come in to parade (*our* parade) which gets through just in time for supper. Immediately after supper, we go to Capt. Swift's and read "sapping and mining" until Tattoo. I then go down to Tattoo, roll call, and my day's work is over. So you see that I am excusable for not writing.

We expect to sail for Mexico in about two weeks, there may be a delay of some few days over that time, but our presence with the Army is so important that we will go at the earliest practicable moment. So we get rid of all the fever business, and will arrive at a very pleasant time.

In order to occupy you a little, I'll give you a few little commissions to execute for me. When I mention a particular number of any article, get no more because you must remember that we don't carry a room and bureau along with us, and I mention the greatest number that I can take. I shall want no more things like those you had made for me, shall even send some home with you, as well as the most of my civilian clothes. I wish two (2) red flannel shirts, two coarse *check* shirts, that blue and white *apron* stuff; four colored shirts—these last to be of cotton (I think, at all events not linen, but check) and of dark colors so as to remain clean for some time—the ordinary colored shirts are what I want. Do see that the buttons are sewed on tight with silk. Remember that these are for service in the field, and must he *cheap,* coarse, and strong. Nothing fine, for Heaven's sake.

I also want four undershirts, of flannel, I suppose—John or Father can tell best of what material these should be. If they ought to be cambric flannel, you need not get any, for I have some now which I have never worn. Two pair red flannel drawers, and four others according to "Doctor's" directions. Do have them made to fit well, and do not let any of them be too large, so as to incommode me. I also want twelve pairs *woolen* socks (not Merino, but regular woolen ones.)

I believe these are all I want. Do try to have them ready by *Saturday next.* Now let me again beg you not to get these of an expensive kind, for that would defeat the very object I have in view. Remember that in active service we go in for the useful and substantive, not for appearances.

I think I shall come on to Philadelphia next Saturday. I may start on Friday, but I should like to take a look at the ball while I am about it. I may not be able to come on before next week, but certainly will come and pay a short visit before we go. I wish you would come on here about next Monday, the board will then have left the Hotel. I will be, comparatively speaking, not quite so busy as I am now, and you will then see our company in its uniform. If you cannot get off, John must come, but do you try to come. I know you will like your visit.

Besides, I want you to come on for the credit of Philadelphia. Those Levis and Miles [Mills?] have "gone on" so that Philadelphia manners are decidedly below par. Do you think—that old wretch Miles accosted me at the dinner table (just as I was sitting by Mrs. Smith) by tapping me on the shoulder, and bellowing out "George" as if he was the best friend in the world, whereas I never spoke to him in my life. He went on talking in that way, but I looked at him, and spoke to him in a way to astonish him a little. He tried the same game the next day, and made some observations which were entirely too familiar and impertinent to suit my taste so I "faced him off" in such a way that I have been "Mr. McClellan" ever since. That youngest daughter has made the greatest fool of herself you can imagine and is the talk and ridicule of everyone on the Point.

They and the other crowd had amused themselves by representing to everyone about their intimacy with *you;* I took the responsibility of correcting such impressions, for I could not imagine that such an intimacy could exist at all, much less without my knowing something about it. Should you come on before they go (they leave after the ball) I hope to see it proved that I was right about *the intimacy.* "Dr. John" it appears is also a very intimate friend of the family!!!!(?)

The course I have pursued towards such persons may be wrong, but I firmly believe that it was right, and intend to follow it out always. Had they acted *decently* even, I should have treated them politely, and shown them some little attention on Father's account, but they thought proper to act otherwise, and I thought it my duty to show them that I was neither a child [n]or a fool.

I received a letter from Frederica which I have already answered. She had received the pictures &c. &c. Also answered my enquiries about the [fortunes?] &c.

We have a very fine company indeed, and expect to "astound the natives" if they will only give us half a chance. I expect that we will have a fine time during this whole campaign. We will probably be back here next summer, altho' the war *may* last longer than we expect.

I have my full dress uniform now. I had not intended to get one yet, but the Captain wished me to do so, and it only cost about $145!

Remember me to Aunt Steinmetz, Judge Coxe, &c. I shall not have time to see any of them when I come on to Philadelphia, without it should be Aunt Steinmetz who lives near us. Do write soon, my dear Mother, to your affectionate son,
George

I want some apple dumplings for dinner on Sunday! I will write a few lines as soon as I am certain about coming on next Saturday. I was in New York for one day this week to get my uniform and went over to see the California Regiment. I was very much amused, you may imagine. Kiss Arthur and Mary for me.

George B. McClellan to John H. B. McClellan

WEST POINT

C. 25 AUGUST 1846

I have only time, my dear John, to scribble a few lines in apology for not having already written to you, but I give you my word that I have not had time to do so. I am kept busy from eight in the morning until dinner time. After dinner I have to study sapping and mining until the afternoon drill, after which I go to parade. After tea we (Capt. Swift, Gustavus, and myself) generally have a consultation; then I go to tattoo. The amount of it is that we have to organize by the 1st of September the first Corps of Engineering Troops that have ever been in this country. The men are perfectly raw, so that we have to drill them, and we are now (today) commencing the practical operations to prepare us for the field. Gustavus and I have been in the woods nearly all the morning with the men, cutting wood for fascines, gabions, &c., &c.

We have now fifty men, and fine ones they are, too. I am perfectly delighted with my duties. The Captain wants a good barber, and a baker, two or three tailors would be not objection. Now if you know of any men who would answer, do drum them up. We have so much to do here, that I cannot possibly be spared to recruit now. I ask you rather than Father because I would rather trust to your judgment in the matter.

Remember that the men will be subject to the ordinary military duties and that the exercise of their trades will be so much extra work for which they will receive better compensation. They *must be fine looking men*, at least 5 ft. 6, and must pass a surgical examination here. They must be intelligent men, and have some education—also Americans.[1]

The pay is Sergeants, $30.00; Corporals, $18.00; 1st Class Private, $13.00; 2nd Class Private, $9.00 per month in money, besides their clothes and rations, so the amount of $12.00 a month. Remember that these clothes and rations are purchased by the wholesale by [the] Government so that they will, probably, get as much for this $12.00 as they would get anywhere else for $18.00 or $20.00.[2]

If you come across any likely men, send them on—but none but good looking, respectable men. We have so many now here, and our company has so high a name, that we can well afford to be particular in receiving men. We do not expect a set of men like the Apollo Belvedere, but *good* looking *men.* Do by all means get the tailor for us.

A very provoking, yet amusing thing came off when the Levis crowd, of which I shall make you full account on Sunday, my first leisure day.

Give my love to Mother and Father, Arthur and Mary. Remember me to Billy Rawle[3] and believe me to be, your affectionate brother,

Geo B. McClellan

P.S. Miner has gone to California. I spent Monday with him in N.Y. and got here on Monday evening.

1. On 19 May 1846 Joseph G. Totten wrote to Alexander J. Swift, informing him that, in relation to recruiting men for his company, "we should be particular in the selection." For engineers he specified, "We must have smart, active, able bodied young men" with a minimum height of five feet six inches, instead of five feet five inches specified by regulation, and a medium height, "from 5' 8" to 5' 10", would be preferred." Moreover, the chief engineer admonished Swift to pay particular attention "to the mental capacity and general intelligence" of the recruit and that each of them "must at least be able to read and write" and "to have learned some mechanical trade." Totten further specified that only single men and native-born citizens were to be enlisted. George B. McClellan Papers, Library of Congress.

2. Military pay, unchanged since 1802, amounted to eleven dollars per month for infantry privates, but sappers and miners could expect an additional twenty-five to forty cents per day. At this time a second lieutenant received twenty-five dollars per month.

3. William Henry Rawle (1823–89) was a prominent Philadelphia attorney.

George B. McClellan to Thomas C. English

WEST POINT

28 AUGUST 1846

Dear Tom,

I write in a dickens of a hurry to know whether *Songo*[1] will go with me to the Wars, as my servant. I can't find a good servant and I concluded to write and enquire whether he would consent to go with me and Lt. Smith of my Company. If you can write back at once without the loss of a single mail, I would be much obliged as we shall sail in about 2 weeks. May be delayed two or three days. I shall not engage anyone until I hear, provided I hear at least a day or so before I sail. As soon as I hear, I will

write back to you. Lt. Smith will go *via* New Orleans in order to take his wife home to Kentucky, and will take my horse on from there, so that if Songo concludes to go with us, he can take my horse to N.O. and then join Smith. As soon as we receive *positive* orders, I will write and tell you exactly when to send my horse to N.O. We have a splendid Company, all Americans, and anxious for a fight. If we don't do something, it will be because they won't give us a chance. I've been on my feet ever since 5 this morning, and now have to start out again immediately to Engineer drill. We got along gloriously with our Ponton Bridge. Have in 2 drills made a bridge in the same time that the French pontoners do (three minutes to a boat), and the French Corps of Pontoners is considered the best in the world and has been in existence for more that 30 years. We will beat them before we are done! Do write to me at once, for a day's delay may cause the letter to reach after I leave, and it is a matter of interest to me, you may imagine. Excuse my brevity, but I'll write you some long ones from the seat of war. I must start right off for drill. Give my love to Sister. Kiss the baby, and believe me to be Your affectionate brother,
Geo B. Mc
Respects to Col. Darrington.

1. Songo was a free black man employed by McClellan's brother-in-law, Thomas C. English. During the campaign in Mexico, Songo rode a horse named Jim Poney that he had brought from English's plantation.

James Stuart[1] to George B. McClellan

FORT McHENRY
1 SEPTEMBER 1846

Dear George,

Having just returned from the city of Washington where I have been on a leave of two days to see and bid farewell to my family, your letter was late in coming to hand. Unfortunately, circumstances, pecuniary and otherwise, prevent my meeting you in Philadelphia, but fortune has favored us in sending us both down to Texas in such fine company, where I hope to have many a hunt and pleasant jaunt with you, my well tried old and

only friend. Our company is off for Brazos Santiago on Friday or at furtherest on Saturday, and I only wish that you were the 2nd Lieutenant of it; you could not be better pleased with a set of men, officers and privates, except perhaps your own. May we both be lucky and come home old soldiers in glory and experiences, but not with [*illegible*] of broken bones and skulls. My hunting suit made expressly for service can resist everything but cannon balls and yellow fever and has more pockets, and bowie knives and revolvers and other murderous implements stowed away in them, than will suffice to kill all the Mexicans we shall ever fight. I should have written to you before, but really have been more busy than at any other time of my life. A large assortment of letters were forwarded to me from Georgetown some time ago, written by you, and if they had only reached me when I had nothing to do, your postage bill would have been much greater. As it is, you must take my word and will for the deed and not leave off writing. Wishing you and yours all good things and anxiously anticipating the long talk we will have together in Mexico, I am dear George,
Your sincere friend
Jas

P.S. By all means keep the *valedictory* for me. This puts me in mind of poor Joe Bacon's death, which is but the beginning of the end.[2] We will all shortly drift off; such being the case, let us sorrow when our classmates are removed, but remember at the same time that it is not in our power to prevent the end. Your sepia is now framed and at home and by far the prettiest thing in the house.

1. James Stuart of South Carolina was graduated in McClellan's West Point class of 1846 and was brevetted to second lieutenant with the Regiment of Mounted Riflemen. He received brevets through captain for "gallant and meritorious conduct" in the battles around Mexico City. Stuart was mortally wounded in action against the Rogue Indians on 17 June 1851 near Rogue River, Oregon. McClellan wrote in his diary: "On the 18th June, at five in the afternoon died Jimmie Stuart, my best and oldest friend. He was mortally wounded the day before by an arrow, whilst gallantly leading a charge against a party of hostile Indians. He is buried at Camp Stuart—about twenty-five miles south of Rogue's River, near the main road, and not far from the base of the Ciskious [i.e., Siskiyous] Mountains. His grave is between two oaks, on the left side of the road, going south, with J. S. cut in the bark of the largest of the oaks."

2. Rufus Joseph Bacon graduated with the West Point class of 1846 and was brevetted into the Fourth Artillery. Before joining his regiment, however, he died on graduation leave in his hometown of Braxton, Maine, 12 August 1846.

George B. McClellan to Thomas C. English

WEST POINT

22 SEPTEMBER 1846

My dear Tom,

I have only time to write a few lines to tell you that we shall certainly sail on the day after tomorrow (the 24th). We have a fine ship, favorable winds prevail, and we have every reason to expect a short and pleasant passage. We sail direct for Point Isabel without touching at New Orleans. Now if you can send my horse to New Orleans to the care of the U.S. *Quartermaster* at that place, requesting him to send the horse on immediately to Pt. Isabel to the care of the U.S. Quartermaster at that place, there to await the arrival of "Lt. Geo. B. McClellan, Company "A" Corps of Engineers," you will confer the greatest possible favor upon me. If you can spare Songo (provided he will consent to go with me) I would be infinitely obliged if you would let him go as servant to Smith G. W. [*sic*] and myself, he will merely have to take care of our horses and "traps." We won't trouble him much personally. As to his wages, tell him we won't quarrel. We will give him any reasonable sum. You can have him again at the close of the war, or I will keep him, just as you please. If you can possibly spare Songo, do let him go, for we will have the very devil to pay if we don't get him. As to my horse, I take it for granted that it is only necessary for me to say that I want a handsome one that can carry me through the hard work of a campaign—pretty stout, of course—all the rest I leave to your judgment, confident that I shall be pleased with your selection.

He had better be sent on as soon as possible after the receipt of this letter. We start with about 75 men—the best Company (so Genl. Scott and Col. Totten both say) in the service, all Americans—all young—all intelligent—all anxious, very eager for the campaign, and above all, well drilled.[1] If the Lord and Santa Anna will only condescend to give us a chance, I'll be most confoundedly mistaken if we don't thrash them "*some.*" Give my love to Frederica and kiss *the* "Nephew." All well at home. The Doctor was up here about two weeks ago, John about one week ago. I'll write a decent *letter* on board ship, and send it from Pt. Isabel—until then goodbye and believe me to be your affectionate brother.

Do write to me in Mexico. Direct to "Lt. Geo. B. McClellan, Company "A," *Engineers,* care of U.S. Quartermaster, New Orleans, La." *Don't* let me be disappointed about the horse and man. I don't know what I can do if I am. Please send enclosed with the horse, and a few lines to U.S. Quarter Master at New Orleans.

1. Joseph G. Totten, chief engineer of the U.S. Army, inspected the company on 19 September 1846, just prior to its departure for Mexico, and found the men to be "intelligent, mostly mechanics"; to have "excellent physical qualities'" and to be "already well set up and disciplined soldiers." The company's drill as infantry "was excellent," he reported to Secretary Marcy, and he found "all the work they had accomplished as sappers to have been executed in a credible manner." The company will, he assured the secretary, "prove of great use, and be distinguished in actual service." Totten to William L. Marcy, "Report of the Chief Engineer," in *Report of the Secretary of War,* 29th Cong., 1st sess., 1846, S. Exec. Doc. 1, 137.

2

"I WOULD NOT HAVE MISSED COMING HERE FOR ANYTHING"

FROM WEST POINT TO VICTORIA

8 October 1846–20 December 1846

At last, on 26 September 1846, the engineer company—seventy-one strong—sailed from New York for Brazos Santiago and service in Taylor's Army of Occupation. Totten referred to the company as being "in admirable discipline, a highly drilled body of infantry and much advanced in their special exercises." McClellan's fellow engineer, Isaac I. Stevens, claimed never to have seen "so superior a company of soldiers." They were, he wrote, "Americans all, young men, having character, zeal, and intelligence, proud of their duties and of their position, perfectly subordinate, and cheerful in their obedience."[1]

Much to McClellan's disappointment, however, the engineers arrived too late to take part in the Battle of Monterrey (20–24 September 1846), in which Taylor soundly defeated the Mexican army. On 11 October the company landed at Brazos Santiago at the mouth of the Rio Grande and marched overland from Matamoras to Camargo, arriving on 2 November. Rather than joining Taylor at Monterrey as planned, however, the company—delayed due to lack of transportation for the heavy engineering train—returned to Matamoras on 29 November.[2]

McClellan missed most of this frustrating marching and counter-marching, making the journey by steamboat. Almost immediately upon landing in Mexico, he (along with nineteen of the enlisted men) had been stricken with dysentery and malaria, confining him to hospital quarters in Camargo for the latter half of October and the first two weeks of November. There too Captain Swift became ill with a fever and was forced to remain at Matamoras when the company, under Smith's command, departed for Tampico on 21 December 1846.[3]

1. Totten to Marcy, "Report of the Chief Engineer," in *Report of the Secretary of War,* 31st Cong., 1st sess., 1849, H. Exec. Doc. 1, 279; Isaac Ingalls Stevens to Father, 11 Apr. 1847, quoted in Hazard Stevens, *The Life of Isaac Ingalls Stevens,* 2 vols. (Boston: Houghton Mifflin, 1900), 1:118. See also William L. Marcy, *Report of the Secretary of War,* 29th Cong., 2d sess., 5 Dec. 1846, H. Exec. Doc. 4, 57.

2. Rodenbough, *Army of the United States,* 116; *Report of the Secretary of War,* 279.

3. McClellan to mother, 14 Nov. 1846, George B. McClellan Papers, Library of Congress.

George B. McClellan to Frederica M. English

SHIP *ORATOR,* GULF OF MEXICO

8 OCTOBER 1846

Here we are, my dear Sister, almost in Mexico. We left West Point on Thursday the 24th (just two weeks ago) and sailed on the 26th. We have thus far been particularly fortunate. We had a very severe gale on the Monday after we left, but it was just "aft." We "loafed" around the "Hole in the Wall"[1] for three days, and have been running under studding sails nearly all the time since. We will, if the wind holds, make Brazos in two and a half or three days. Altho' it is by no means impossible that calms, head winds, &c. may keep us a week or so.

You must remember the Hole in the Wall, without you passed it in the night. We contemplated it at our leisure for three days, and I fancy that it is tolerably well impressed on my mind. I have nothing more to tell you about our voyage, for you have been over pretty much the same track as far as the Tortugas, and our voyage has been, I suppose like all others, except that we have been very lucky so far.

As a matter of course, we have seen sharks, nautilus, dolphins, and thousands of flying fish. I suppose you saw the same. I have been very fortunate as to seasickness, also.

I received your letter a day or two before we left West Point and was, I assure you, very much delighted to hear that the servant [Songo] and horse would be on hand. You ask me whether I carry only a sword. Indeed, I don't. I have a double barreled shot gun, for game; a bowie knife; two revolvers; a sabre, and a rapier—the last only for dress occasions, the others for use. Our men act as Infantry. The drivers of the wagons are furnished by the Quarter Master's Department, and are *under our control,* tho' not belonging to the Company.

As for the Pontons, they are made of India rubber and inflated with air. Each ponton is formed of three cylinders, eighteen feet long and nineteen inches in diameter. These cylinders are separate internally, but connected together externally by the India Rubber cloth along the "elements of tangency." They are pointed at the ends, and the three together form one boat. The cylinders are inflated separately by means of an ordinary bellows applied to an opening in one end, this opening being provided with a valve. This boat is much better than a wooden boat. It draws less water, is very much lighter, and is much more easily managed. I'll try to give you some kind of an idea of the bridge.

Brazos Santiago
16 October 1846

I found that I had to give up as a bad job the idea of giving you any conception of our bridge on board ship, and my conveniences are now so limited that it is a still more hopeless case. I am now on guard, writing in the guard tent, a chest for a desk, the sand for a seat, the sand for a prospect. We have encamped on a sand bar about nine miles long and one half mile broad. There is nothing but sand to be seen in any direction, not the slightest shadow of vegetation, there being in some favorable places some three blades of grass to fifty square feet of sand.

Whenever the wind blows, the sand comes along in perfect clouds, filling your tents, eyes and everything else. I am not at all disgusted or discouraged, except with the fact of that confounded battle taking place before we could by any possibility participate in it.[2] It is to us a piece of bad luck, which I shall regret as long as I live. The only consolation is that the fellows may be "cussed fools" enough to give us another chance at them at Saltillo. If they do, look out for Co. "A" Engineers. We will put in the licks for four battles at once.

I have not as yet heard anything of the horse. I suppose it will come in good time. This is the 16th. I shall have seen several officers who were in the battle, three of them dined with us today. We will leave these diggings for Matamoras early tomorrow morning. I will write to you from thence all about the country, the city &c. &c.

As I [am on] guard I can write no more just at present and as it is probable that I shall not have another opportunity of writing tonight, I will now say goodbye until we get to Matamoras. This may not reach you for a month, the mail is so uncertain. Direct to "Lieut. G. B. McClellan, Comp. "A," Engineers, with the Army of Mexico at Monterrey, via Brazos Santiago." Love to Tom and the Nephew and believe me
Your affectionate brother
Geo B McClellan

1. The Hole in the Wall is a natural arch at the south end of Abaco Island in the northeastern Bahamas. According to one voyager, "through this large arch the sea generally breaks, and thus forms a conspicuous landmark for navigators that pass this island." George Coggeshall, *Thirty-six Voyages to Various Parts of the World* (New York: George Putnam, 1858), 371.

2. Zachary Taylor's army followed the Army of the North, now under the command of Gen. of Division Pedro Ampudia, from the Rio Grande deep into Mexican territory, where in three days of fighting, 21–23 September 1846, the Americans captured the fortified city of Monterrey.

Diary Entry

BRAZOS SANTIAGO, TEXAS

24 SEPTEMBER 1846–10 OCTOBER 1846

We left West Point on the 24th of September 1846 for General Taylor's army in Mexico—Company "A," Engineers consisted of Captain [A. J.] Swift, Lieutenant G. W. Smith, myself and seventy-one rank and file. On Saturday the 26th we sailed from the Narrows bound to Brazos de Santiago where we were so fortunate as to arrive in fourteen days. We had a very pleasant passage, on the whole. Felt very much the want of ice, and claret. At one time could only eat raw tomatoes.

The result of my experience with respect to the transportation of troops by sea is—In the first place see that the part of the vessel destined to receive them is thoroughly policed, washed and well scraped out before the vessel sails; then let a strong police party be detailed every day, so that the part between decks may always be well washed out and smell well. Windsails are very necessary. The acting commissary of subsistence should see for himself exactly what is put on board for the use of the troops and should cause a written requisition to be made upon him for the quantity used from day to day or week to week. He should have a reliable and intelligent sergeant at his disposal. Care should be taken that good cooking arrangements are provided. Mush appeared to be a favorite and agreeable food for the men at sea. The muskets should be inspected every day, when the weather permits, as also the quarters. Men must be required to wear their worst clothes (working over-alls, etc.) on board. Care should be taken that camp equipage and all articles necessary for immediate use of troops when landed are so stowed that they can be got at at once.

Brazos is probably the very worst port that could be found on the whole American coast. We are encamped on an island which is nothing more than a sand bar, perfectly barren, utterly destitute of any sign of vegetation. It is about six miles long and one-half mile broad. We are placed about one hundred yards from the sea, a row of sand hills some twenty feet high intervening; whenever a strong breeze blows the sand flies along in perfect clouds, filling your tent, eyes and everything else. To dry ink you have merely to dip your paper in the sand. The only good thing about the place is the bathing in the surf.

The water which we drink is obtained by digging a hole large enough to contain a barrel. In this is placed a bottomless barrel in which the water collects. You must dig until you find water, then "work-in" the barrel until it is well down. This water is very bad. It is brackish and unhealthy.

The island is often overflowed to the depth of one or two feet. To reach this interesting spot, one is taken from the vessel in a steamboat and taken over a bar on which the water is six feet deep, and where the surf breaks with the greatest violence. It is often impossible to communicate with the vessels outside for ten days or two weeks at a time.

Diary Entry

BRAZOS SANTIAGO, TEXAS

14 OCTOBER 1846

We have been here since Monday afternoon [10 October] and it is now Friday [14 October]. We expect to march for the mouth of the Rio Grande tomorrow morning at break of day—thence by steamboat to Matamoras where we will remain until our arrangements for the ponton train are complete. We received when we arrived the news of the battle of Monterrey. Three officers who were present dined with us today—Nichols of the 2nd Artillery,[1] Captain Smith (brother of G. W. Smith) formerly Captain of Louisiana Volunteers now an amateur,[2] Captain Crump of the Mississippi Volunteers[3]—fine fellows all. Saw Bailie Peyton[4] and some others pass our encampment this morning from Monterrey.

I am now writing in the guard tent (I go on guard every other day). Immediately in front are sand hills, same on the right, same in the rear, sandy plain on the left. To the left of the sand hills in front are a number of wagons parked, to the left of them a pound containing about 200 mules, to the left and in front of that about fifty sloops, schooners, brigs and steamboats; to the left of that and three miles from us may be seen Point Isabel.

1. William Augustus Nichols was graduated with the West Point class of 1838 and was commissioned into the Second Artillery. He was twice brevetted for his conduct during the Mexican War.

2. John Rhodes Smith, the elder brother of Gustavus Woodson Smith, was elected a captain in the Fourth Louisiana Infantry but was discharged, with the regiment, in August 1846 before it was deployed to Mexico.

3. George A. Crump of Vicksburg was the captain of Company H of Col. Jefferson Davis's First Mississippi Rifles. He resigned from the regiment at Victoria on 21 December 1846. Joseph E. Chance, *Jefferson Davis's Mexican War Regiment* (Jackson: University Press of Mississippi, 1991).

4. Bailie Peyton was a member of the U.S. House of Representatives from Tennessee and later the U.S. district attorney in New Orleans. In 1841 he declined appointment as secretary of war in the John Tyler administration. At the outbreak of the Mexican War, he was elected colonel of the Fifth Louisiana Infantry, but when James K. Polk refused the service of the regiment, Peyton served as a volunteer aide, first to William J. Worth and then to Zachary Taylor.

George B. McClellan to Elizabeth Brinton McClellan

CAMP OPPOSITE CAMARGO, MEXICO

14 NOVEMBER 1846

You must think me a perfect jewel in the line of writing letters, my dear Mother, but now that the dog is dead, I will tell you the reason why. While at Brazos and on the way thence to Matamoras, I was too busy to write; when I arrived at Matamoras I was almost immediately taken sick. I remained so during the two weeks we stayed there, and whilst on the steamboat thence to Camargo. When we got here, I went into Hospital Quarters where I emerged yesterday, so that I have had almost a month's sickness, but now I am perfectly well and "ready, aye ready for the field."[1]

My quarters in Camargo were large apartments in palace of Don Jesus Guiste, brother of the Alcalde. The Don has left for parts unknown. I wish you could see the gorgeous paintings on the walls of the palace; you would laugh your eyes out, and *then* you wouldn't stop. Mexico is altogether the queerest place I ever came across—it is chaparral, chaparral, nothing but chaparral so far. The people are very polite to the Regulars (Soldados spéciales de la leina) but they hate the Volunteers as they do Old Scratch himself. Their dress is very pretty; quite picturesque, indeed. The men are very fine looking, rather superior to the corresponding sex of our people. The women, as far as I have seen, are a very dirty, ugly looking set. No black-eyed Senoritas have as yet been visible.

The generality of the houses are miserable huts, built of upright logs, canes and mud, with thatched roofs. If you look in one, you will see in the simple apartment about a dozen men, women, children, dogs, goats, &c., &c. in the dirt. They are a very dirty people in their houses. Some of the houses (in the towns) are built of stone with flat roofs. They seldom have windows, and when they do, they are closed by a grating. Glass is unknown amongst them.

I rode out this morning with Jimmy Stuart to a very pretty rancho, indeed, some of the houses built of stone. The people were rather sulky—probably because the volunteers are very troublesome to them. A Mexican was shot a mile or two from here by some volunteers yesterday afternoon. This is by no means an uncommon occurrence. The worst thing we will hear will be that a volunteer is shot by way of retribution. You never hear of a Mexican being murdered by a regular, or a regular by a Mexican. The volunteers carry on in the most shameful and disgraceful manner.

The Mexicans are a most deceitful and rascally set of devils. I believe that it is nothing but fear which makes them pay such great respect to the regulars. They would like to cut all our throats in the dark—but, thank Heaven! they are afraid. Should we meet with reverses, and be forced to retreat, not a straggler would escape. Our safety lies in our success.

As I mentioned just now, Jimmy Stuart is here. He is in my tent now, writing at the same table. He has been here about a month, and leaves this afternoon or tomorrow morning for Monterrey with his Company. It was very fortunate for me that I found him here, for he was of course very kind to me, in taking care of me when I was sick.

Smith and myself are alone here, the Captain having gone down to Matamoras on public business. The ponton train is to be collected at Matamoras. We wait here until further orders. As soon as it is ascertained whether there is a passable road from Matamoras to Tampico, it will be determined whether Genl. Patterson's column moves by that road or by sea.[2] If he goes by land, we will go with him, if by sea, we accompany Genl. Taylor. All that I want is a move somewhere. There can be no fight until Tampico is reached, so that we shall miss nothing.

There is a rumor going about that Tampico has been taken by the Navy.[3] If that is the case, we shall have no fight before reaching San Luis Potosi.

The nights are becoming most delightfully cool and the days not *quite* so hot. Does it not appear strange to you, that whilst you are shivering in the middle of November, we are roasting in our tents? This is a most barren and uninteresting country—not entirely uninteresting either, for I find a great deal of amusement in observing the habits of the people, &c., &c. I would not have missed coming here for anything in the world, now that I am well and recovering my strength. I commence to enjoy the novelty of the affair, and shall have enough to tell you when I return, to fill a dozen books. The only trophies I have picked up yet are some buttons from a dead Mex's coat on Palo Alto, some bullets and buttons from Resaca.

Talking of Palo Alto reminds me that I killed an enormous rattlesnake there. I shot him through the head with a horse pistol, just as he raised his coil. I cut off his rattle for Arthur, but it got lost at Matamoras. Never mind. I will kill plenty more, I reckon. Tell John that if I ever go to Palo Alto again, I will try to bring a Mex skull; there are plenty of them there, but it is very hard to carry anything. I should not be surprised at any time to have to throw away my trunk.

I have my horse safe and sound, together with Songo. It is the Duane mare. She is a noble animal, said to be the best in Mexico. The only thing is that she is too good for this infernal country. We live a pretty hard sort of a life, but not so bad *yet* as you might imagine. It is a glorious life, but nevertheless I shall enjoy the comforts of civilized life most amazingly when I get back. I would not for the world return before the end of the war, or without seeing a fight, but I shall by no means be displeased if both of these desirable objects are attained by next spring. I should be willing to crowd all the rest of the fighting that has to come off into a month's space, and then make tracks for the "States."

We have very nice oranges here, but they (as everything else) are very dear. The nearer you get to Monterrey (where they grow in great abundance) the more they ask for them. These Mexicans are a precious set of cheats. The scamps will cheat you out of your eyes if you are not careful. The only way is to bully them a little.

Smith is very well and sends his regards. Stuart ditto.

They have three bells in the church here and they manage to keep up the most horrible racket you ever heard in your life. It almost set me crazy when I was on the other side. They are at it now, full tilt. They say that whenever the Priest gets out of funds, he rings up his parishioners. He must have been in very reduced circumstances ever since I have been here. He's poor as a rat today.

A Mexican funeral is really ludicrous. A man dressed in a red frock runs before the procession with a cross in his hand, and the rest follow at a dog trot. There is nothing whatever of the solemn in it. Whilst performing mass, they start a clarinet, which plays tunes sounding mightily like Lucy Neal &c. &c.[4] Certainly as unlike sacred music as possible.

Now is our dinner time and I must stop. If no mail leaves for a day or two, I'll write another letter and send it in the same envelope.

Hurrah! Just made a raise in some oranges—going to get some ginger bread this afternoon. Won't we have a supper!

Sunday morning. The Mail leaves this morning, so I will send this at once and write by next mail.
Your affectionate son
Geo B McC[5]

1. McClellan here quotes from Sir Walter Scott's epic poem, *The Lay of the Last Minstrel.* The allusion is to Sir John Scott of Thirlestane, a follower of James V of Scotland in the border wars against England.

2. Robert Patterson served as a captain in the War of 1812, after which he became a wealthy mill owner and active in Democratic politics in Pennsylvania. At the outbreak of the Mexican War, due largely to his political influence, he was commissioned a major general and placed in command of the volunteer division. Although the Patterson and McClellan families were well acquainted with one another in Philadelphia, McClellan felt only contempt for "the old Mustang" as a commander.

3. On 15 November 1846 the Home Squadron under Commodore David Conner occupied Tampico, Mexico's second largest port, unopposed.

4. "Lucy Neal," the story of two star-crossed lovers on an Alabama plantation, was a minstrel song, written in 1844 by James Sanford and popularized by the Ethiopian Serenaders.

5. Before forwarding this letter to his sister, Frederica, McClellan's mother wrote across it: "I will send you this letter of George's because you may not hear from him very soon again as he must be on his way to Tampico now. . . . The remains of Major [Samuel] Ringgold arrived in Baltimore yesterday. There is to be a great public funeral. A 2nd Regiment is raising here and you hear the boys all singing songs about going off to Mexico. You will see that George does not consider it an Elysium. I shall be heartily glad when we get him back."

Diary Entry

CAMP OPPOSITE CAMARGO
15 NOVEMBER 1846

We marched from Brazos to the mouth of the Rio Grande and on arriving there found ourselves without tents, provisions or working utensils, a cold Norther blowing all the time. We, however, procured what we needed from the Quarter Master and made the men comfortable until the arrival of Captain Swift with the wagons, who reached the mouth late in the afternoon, whilst we got there about 10 A.M. Thanks to Churchill's[1] kindness G. W. Smith and myself got along very well.

We left in the *Corvette*[2] the next morning (Sunday) for Matamoras, where we arrived at about 5 P.M. The Rio Grande is a very narrow, muddy stream. The channel is very uncertain, changing from day to day. The banks are covered with the mesquite trees, canes, cabbage trees, etc. The ranchos are rather sparse, but some of them are very prettily situated. They all consist of miserable huts built of mesquite logs and canes placed upright—the interstices filled with mud. The roofs are thatched, either with canes or the leaves of the cabbage tree (a species of palmetto).

Cotton appears to grow quite plentifully on the banks, but is not cultivated at all. The Mexicans appear to cultivate nothing whatever but a little Indian corn (maize). They are certainly the laziest people in existence—living in a rich and fertile country (the banks of the river at least) they are content to roll in the mud, eat their horrible beef and tortillas and dance all night at their fandangos. This appears to be the character of the Mexicans as far as I have seen, but they will probably improve as we proceed further in the country.

Matamoras is situated about a quarter mile from the river. Some of the houses on the principal streets are of stone, there is one near the Plaza built in the American style with three stories and garrets. All the rest are regular Mexican. On the Plaza is an unfinished cathedral, commenced on a grand scale, but unfinished from a want of funds. The great majority of the houses are of log. The place is quite Americanized by our army and the usual train of sutlers, etc., etc. You can get almost everything you want there.

We were encamped near the landing. I rode over to Resaca and Palo Alto, but as there is just now a prospect of our returning to Matamoras, before moving on Tampico, I shall write no description of the fields until I have visited them again.

After being sick for nearly two weeks in Matamoras I left with the company for Camargo on the *Whiteville*,[3] where we arrived two weeks ago tomorrow, and I have been in Hospital Quarters ever since until day before yesterday.[4]

Now I am in camp, the wind blowing the dust in such perfect clouds that it is perfectly horrible—one can hardly live through it. My quarters in Camargo were the Palace of Don Jesus, the brother of the Alcalde [mayor] of the town—he (the Don) having absquatatated [*sic*]. The main body of the Palace (!) is one storied. It consists of two rooms—the smaller one occupied by Dr. Turner, the other by "Legs" [i.e., George S. Humphreys] and, myself (together with J[ames] S[tuart] for a part of the time). The floor is of hard earth, the walls white, and very fancifully decorated with paintings—the roof flat and painted green—an inscription on it showing that "Se acabó esta casa en este Dia &c. &c. 1829."[5] Altogether it was quite a recherché establishment. Jimmie Stuart came down to take care of me when I first got there, and after doing so with his usual kindness was unfortunately taken with a fever and had to stay there anyhow.

We are to accompany General Patterson to Tampico. I hope and suppose that we will have a fight there, then join General Taylor, then hey for San Luis [Potosi] and another fight.

1. William Hunter Churchill, class of 1840, was commissioned into the Third Artillery and brevetted to captain for "gallant and distinguished services" at the battles of Palo Alto and Resaca de la Palma. He was assigned to duty as quartermaster at Fort Polk, Point Isabel, Texas, where he died on 19 October 1847.

2. The side-wheel steamer *Corvette* was purchased by the U.S. Quartermaster Department to transport troops, stores, and equipment from Brazos Santiago to Fort Brown. Due to the boat's reputation for speed and comfort, Winfield Scott chose it to transport him and his staff to Camargo in January 1847.

3. The *Whiteville* was one of five light-draft steamers purchased for the army's campaign up the Rio Grande. *Martinsburg ([W.] Va.) Gazette,* 23 July 1846, 2.

4. Upon learning of his illness, Dr. George McClellan advised his son to "take a grand purge now and then." "Clear the stomach of all undigested matter the moment you feel sick," he prescribed, "by drinking copious draughts of warm water and if necessary a teaspoon full of mustard infused into every tumbler. The Irishman's emetic (the fingers down the throat) will do in case every thing else fails." Dr. George McClellan to George B. McClellan, 19 Mar. 1847, George B. McClellan Papers, Library of Congress.

5. "This house was completed [on a certain date] in 1829."

George B. McClellan to Charles S. Stewart[1]

STEAMER *CORVETTE* ON THE RIO GRANDE

1 DECEMBER 1846

My dear Captn

I am afraid I have run up a tremendous score of demerits for lates &c &c. since your last letter, for I suppose you have considered yourself at perfect liberty to poke it into me for not answering your last letter, but I don't care, it won't do Blunt[2] any good now! The fact is that during the month preceding our departure from West Point we were so very busy that I could do nothing in the line of writing letters. I only went home for *one* day to say good bye.

We had a very short passage and as pleasant I suppose as such things usually are when done up in a merchant ship with poor accommodations and poor fare. We landed at Brazos de Santiago (that's the very best Mexican) on the 12th of October and encamped there for one week, where Smith and myself went on guard every other day and sat up during the whole of the "night belonging thereto" without closing an eye—for we had a perfectly raw set of noncommissioned officers and men who did not know what a sentinel was. The only thing that kept us alive was the sea bathing.

The water there is horrible. It is brackish and very unhealthy. The place itself is a flat sand bar, destitute of vegetation, and whenever the wind blows it carries along with it clouds of sand which fills the eyes, clothes, and everything.

From this place to the mouth is a march of 10 miles on the sea beach; here you take the steamboat. Went up to Matamoras where we landed. It is about 100 miles to that place by water, the river being very remarkably torturous. The current of this river is rapid and the water very muddy, the banks very monotonous, but abounding with game.

We remained over two weeks at Matamoras where your humble very shortly caught the dysentery and has been sick now for somewhat more than a month. I have also had the chills and fever, but am almost well of both. Of course, I have not had the *real* dysentery *all* the time, but it took a more robust form.

We then went up to Camargo where we received orders from Genl. Taylor to report to Genl. Patterson. It was then reported that Tampico was strongly fortified and garrisoned and that we would have a grand fight there—so I set to work with great goût to draw maps of the route to Tampico, plans of Tampico and its fortifications, and was in fine spirits, when slap dash go all our high hopes upon the arrival of the news that the Navy had taken the place without firing a single gun. We are now on the way (that is, we would be if the wind was not so high) to Tampico where a force of 5000 men will be collected in the course of two or three weeks under the old Mustang (Patterson).

We will (that is, the *Engineers*) repair the fortifications, and the communications between it and Monterrey will then be established. One point of the line (Matamoras) is already occupied by the 2nd Infantry. You see that we will then have in our possession all the country between the Rio Grande and that line. We will, in the meantime, have collected large supplies at Tampico, both of ammunition and provisions. The new levies (which are absolutely necessary if the war is to last) will collect there and then we will be ready to make a decisive movement.

Several cases may arise. We may occupy the extent of country I just mentioned and make an attack on Vera Cruz, moving from Tampico by sea, thence to the Capital, disregarding San Luis, or we may move on San Luis both from Monterrey (which is the best road) and Tampico and then on Mexico. Heaven alone knows what they intend to do—I suppose we will find out when the movement is finished, if not before.

Matamoras is quite a nice place (*for Mexico*). There are many stores there now kept by Americans so that you can procure many conveniences there. The best houses are built of stone. They are one storied, with flat roofs, have one door, one window, without any glass and look more like stables or prisons than private houses.

The ranchos are all built of Mosquite [*sic*] logs and cane. They are without exception the most uncomfortable, the dirtiest, most Mexican things you can imagine. There they crowd together men, women, children, pigs, goats, and everything else. No modesty, no decency, and no cleanliness.

Camargo is much worse. It has the peculiarly pleasant characteristic of being the dustiest place on the face of the Earth. Your eyes and everything you have must be filled with dust there and the water is bad. I was for nearly two weeks in quarters in this town (since I was sick). I was in the same room with Legs Humphreys.[3] It was a portion of the *Palace* of Don Jesus Guiste, the Alcalde's brother—a pretty specimen of a place it was, too.

Legs has gone down to Matamoras on his way home. He has been ordered out of the country by the Surgeon on account of his health; the surgeon tells me that he never can do duty in this country, especially in the Dragoons, and that he is afraid that he is very apt to live but a very few years, any how.

Neighbor Jones is at Camargo.[4] He has been suffering severely from an injury he received on his heel and from sickness. He is nearly well now and expects to leave for Matamoras with the rest of his regiment in a few days. The two companies of Mounted Rifles were here with the 2nd Infantry until two weeks last Thursday when they left for Monterrey. Rhett, Maury, [and]

Stuart were attached to them. Bob Easley has been sick at Camargo but has now gone on to Monterrey. When at Matamoras, I saw Wilkins, Ehninger, Pickett, Sturgis, and Easley who were passing up. I hear that Botts and Tyler are below. Brown and Sears have gone on to Tampico, also Shields, and, I strongly suspect, the never-sufficiently-to-be-admired Magilton, also.[5]

I am very anxious to get down to the mouth where I'll have sea bathing and fresh mullet in abundance. Tampico will probably be a more pleasant place than any I have yet seen in Mexico. I hope to heaven it proves to be so, for I have not seen much fun out here yet. Never mind. When I once get well, perfectly well, I intend to have my own amusement out of the country. Do write whenever you have leisure and believe me your sincere friend. Geo. B. McClellan

[P.S. in margins] Thursday—We are getting along very well. Will be at Matamoras tomorrow morning. I saw a really handsome senorita this morning. It is a Rara Avis here, I assure you. Notwithstanding all the nice tales and romances about black eyed senoritas, orange groves, and all such damned humbugs.

Give my respects to Capt. Cullum.[6] Tell him of our countermarch.

When you see Mrs. Smith, present my best respects and tell her that G. W. is very well and in good spirits.

Lt. Scarritt has left the country on a sick leave, for two months.[7]

1. Charles Seaforth Stewart was graduated first in the West Point class of 1846—one step above McClellan—and was commissioned a second lieutenant of engineers. He retired from the service on 16 September 1886 with the rank of colonel, having spent his entire career in engineering duties. This letter is part of the Charles Stewart Collection in the manuscript division of the Houghton Library at Harvard University and was generously shared with the editor by Stephen W. Sears.

2. Charles Edward Blunt finished third in the class of 1846, just behind Stewart and McClellan, and was brevetted to second lieutenant of engineers. He saw no field service in either the Mexican War or the Civil War and retired with the rank of colonel on 10 January 1887.

3. George S. Humphreys, known as "Legs" to his fellow cadets, graduated in McClellan's class of 1846. True to the surgeon's prediction, Humphreys died on 9 November 1847.

4. David Rumph "Neighbor" Jones graduated with McClellan. He received a brevet to first lieutenant in Mexico but resigned from the U.S. Army in 1861 to accept a position as chief of staff to P. G. T. Beauregard during the Fort Sumter crisis. Jones ably commanded a division against McClellan during the Peninsula Campaign and at Sharpsburg but died at Richmond on 15 January 1863 of heart disease.

5. Thomas Grimke Rhett, class of 1845, served in the Regiment of Mounted Riflemen during the Mexican War. He was brevetted to captain for his role in the defense of Puebla.

The remainder of the officers in this list graduated with McClellan in the class of 1846. Dabney Herndon Maury served with the Mounted Riflemen and the Third Artillery and was brevetted to first lieutenant for gallantry at Cerro Gordo. His memoir, *Recollections of a Virginian* (New York: Charles Scribner's Sons, 1894), is one of the finest reminiscences of Mexican War and Civil War experiences. "Bob" Easley would seem to be Thomas Easley, who was killed in action at the Battle of Churubusco while serving with the Eighth Infantry. John Darragh Wilkins was assigned to the Fourth Infantry and won a brevet to first lieutenant for gallantry at Contreras and Churubusco. Henry Astor Ehninger served in the Fourth Artillery without particular distinction during the Mexican War. George Edward Pickett, the "goat" of the class of 1846, was assigned to the Eighth Infantry and won brevets through captain for gallantry first at Contreras and Churubusco and then at Chapultepec. He is of course best known as the leader of the gallant, doomed assault on the center of the Union line at Gettysburg, "Pickett's Charge." Samuel Davis Sturgis, Second Dragoons, had an undistinguished Mexican War record but gained some merit in the Civil War, becoming a brigadier general of volunteers. After the war he was appointed colonel of the Seventh Cavalry but was on detached duty at the time of the Battle of the Little Bighorn, 25 June 1876. Archibald Blair Botts, Fourth Infantry, died in Mexico, 1 January 1847. William Henry Tyler, Fifth Infantry, received a brevet to first lieutenant for gallantry at Contreras and Churubusco. John A. Brown, Fourth Artillery, resigned in 1861 with the rank of captain to serve the Confederacy as a colonel of artillery in Robert E. Lee's Army of Northern Virginia. Henry B. Sears, Second Artillery, was brevetted to first lieutenant for his role in the Battle of Cerro Gordo. Hamilton Leroy Shields, Second Artillery, received brevets through captain for his conduct in the battles around Mexico City. Albert Lewis Magilton, Fourth Artillery, received a brevet to first lieutenant for his conduct at Contreras and Churubusco.

6. George Washington Cullum was graduated third in the class of 1833 and assigned to the engineers. He was an instructor at the academy during the Mexican War and was assigned to command of the company of sappers and miners upon its return to West Point, thus for a time becoming McClellan's commanding officer.

7. Jeremiah Mason Scarritt, class of 1834, was a first lieutenant of engineers on Zachary Taylor's staff. He was brevetted to captain for his conduct at the Battle of Monterrey.

Diary Entry

MOUTH OF THE RIO GRANDE

5 DECEMBER 1846

After getting up quite an excitement about a fight at Tampico, etc., we were completely floored by the news that the Navy had taken it without firing a single gun—the place having been abandoned by the Mexican troops, who are doubtless being concentrated at San Luis Potosi in anticipation of a grand attack on the place—ah! if we only fool them by taking Vera Cruz and its castle, and then march on the capital—we would have them completely.

After a great many orders and counter orders we have at length arrived thus far on our way to Tampico. We left Camargo on Sunday evening last (November 29th) in the *Corvette,* with Generals Patterson and Pillow[1] and a number of other officers (among them Captain Hunter, 2nd Dragoons,[2] Major Abercrombie,[3] Captain Winship,[4] Seth Williams,[5] and about a thousand volunteers). We had decidedly a bad passage—running on sand bars very often—being blown up against a bank by the wind—breaking the rudder twice, etc., etc. We left General Patterson, Captain Swift, and many others at Matamoras.

The General started with the intention of going to Tampico by sea—all the troops (except the Tennessee cavalry) were to go by sea, but at Reynosa an express overtook us ordering the General to proceed by land with all the troops except this company, which is to go by sea(!). Captain Swift remained at Matamoras on account of his health.

I was perfectly disgusted coming down the river. I found that every confounded Voluntario in the "Continental Army" ranked me—to be ranked and put aside for a soldier of yesterday, a miserable thing with buttons on it, that knows nothing whatever, is indeed too hard a case. I have pretty much made up my mind that if I cannot increase my rank in this war, I shall resign shortly after the close of it. I cannot stand the idea of being a Second Lieutenant all my life. I have learned some valuable lessons in this war. I am (I hope and believe) pretty well cured of castle building. I came down here with high hopes, with pleasing anticipations of distinction, of being in hard fought battles and acquiring a name and reputation as a stepping stone to a still greater eminence in some future and greater war.

I felt that if I could have a chance I could do something; but what has been the result—the real state of the case? The first thing that greeted my ears upon arriving off Brazos was the news of the battle of Monterrey—the place of all others where this Company and its officers would have had an ample field for distinction. There was a grand miss, but, thank heaven, it could not possibly have been avoided by us.

Well, since then we have been dodging about—waiting a week here—

two weeks there for the ponton train—a month in the dirt somewhere else—doing nothing—half the company sick—have been sick myself for more than a month and a half—and here we are going to Tampico.

What will be the next thing it is impossible to guess at. We may go to San Luis—we may go to Vera Cruz—we may go home from Tampico, we may see a fight, or a dozen of them—or we may not see a shot fired. I have made up my mind to act the philosopher—to take things as they come and not worry my head about the future—to try to get perfectly well—and above all things to see as much fun as I can "scare up" in the country.

I have seen more suffering since I came out here than I could have imagined to exist. It is really awful. I allude to the sufferings of the Volunteers. They literally die like dogs. Were it all known in the States, there would be no more hue and cry against the Army, all would be willing to have so large a regular army that we could dispense entirely with the volunteer system. The suffering among the Regulars is comparatively trifling, for their officers know their duty and take good care of the men.

I have also come to the conclusion that the Quartermaster's Department is most woefully conducted—never trust anything to that Department which you can do for yourself. If you need horses for your trains, etc., carry them with you. As to provisions (for private use) get as much as possible from the Commissaries—you get things from them at one-half the price you pay sutlers.

Smith has ridden over to Brazos de Santiago to endeavor to make arrangements for our immediate transportation to Tampico. Captain Hunter went with him on my mare. They return in the morning. Whilst at Camargo, Smith had a discussion with General Patterson about his (General Patterson's) right to order us when en route to join General Taylor, under orders from Head Quarters at Washington. The General was obliged to succumb and admit the truth of the principle "That an officer of Engineers is not subject to the orders of every superior officer, but only to those of his immediate chief, and that General (or other high officer) to whom he may be ordered to report for duty."

There goes Gerber with his tattoo—so I must stop for the present.[6]

1. Gideon Johnson Pillow was arguably the most incompetent general officer in the American military history. His partnership in the practice of law with future president

James K. Polk seems to have been his sole qualification for high command. With the outbreak of the Mexican War, Polk appointed him a brigadier general and soon promoted him to major general. McClellan wrote in his diary, "M. le President a fait ce que le bon Dieu ne peut pas" (the president has made what the Lord did not), that is, a general of Gideon Pillow. Nathaniel Cheairs Hughes Jr. and Roy P. Stonesifer, *The Life and Wars of Gideon J. Pillow* (Chapel Hill: University of North Carolina Press, 1993).

2. Nathaniel Wyche Hunter was an 1833 graduate of West Point.

3. John Joseph Abercrombie, class of 1822, served as aide de camp to his father-in-law, General Patterson. He was brevetted to lieutenant colonel in the First Infantry for gallantry at the Battle of Monterrey, during which he was wounded. Abercrombie served on the Peninsula as brigadier general of volunteers in McClellan's Army of the Potomac and was wounded at Seven Pines.

4. Oscar Fingal Winship, class of 1836, served as assistant adjutant general of Gideon Pillow's brigade and from 1847 until 1848 on the staff of Franklin Pierce. He was brevetted to captain in the Second Dragoons for his service during the battles of Palo Alto and Resaca de la Palma and to major for his service at the Battle of Churubusco.

5. Seth Williams, class of 1842, was a lieutenant in the First Artillery and an aide de camp to Robert Patterson when he received a brevet to captain "for gallant and meritorious conduct" during the Battle of Cerro Gordo.

6. The company bugler, Cpl. Frederick W. Gerber, a musician in civilian life, was promoted to sergeant to date from 1 February 1849. G. W. Smith cited Gerber's "particularly distinguished" conduct during the final battle for Mexico City. Company muster role; Joseph G. Totten, Engineer Order No. 3, 2 Apr. 1849; John L. Smith to Scott, 26 Sept. 1847, in *Report of the Secretary of War*, 30th Cong., 1st sess., 1848, S. Exec. Doc. 1, 430.

Diary Entry

MOUTH OF THE RIO GRANDE

6 DECEMBER 1846

Go it Weathercocks! Received an order from Major McCall[1] this morning to go back to Matamoras, as we are to *march* to Tampico, via Victoria, with the column under General Patterson. Smith is away at Brazos and if the order had been one day and a half later we would have been off to Tampico by sea.

I have fine sea bathing here. It is blowing very hard from the south east, so much so as to raise the sand too much for comfort entirely.

Bee[2] and Ward[3] at the Brazos—coming over this morning—will at least have an opportunity of giving Georgie [George Horatio Darby?] that letter of Madame Scott's!

I feel pleased at the idea of going by land—we will have a march to talk about, and may very probably have a fight on the way. I firmly believe that we will have a brush before reaching Tampico. Unfortunately the whole column is Voluntario.[4]

1. George Archibald McCall, class of 1822, received brevets through the rank of lieutenant colonel in the Third Infantry for "gallant and meritorious conduct" during the Mexican War.

2. Bernard Elliott Bee, class of 1846, was commissioned into the Third Infantry and received brevets through captain "for gallant and meritorious service" during the Battle of Cerro Gordo, where he was wounded, and the Battle of Chapultepec. As a brigadier general in the Confederate army, he was killed in action at the First Battle of Manassas only minutes after famously dubbing Brig. Gen. Thomas J. Jackson "Stonewall."

3. James Nobel Ward, class of 1845, received a brevet to first lieutenant in the Third Infantry for meritorious service during the Battle of Cerro Gordo, where he was wounded.

4. Patterson's division on the march to Tampico consisted of two brigades of volunteers, those of John A. Quitman and Gideon Pillow. Pillow's brigade was made up of Col. Ferris Foreman's Third and Col. Edward D. Baker's Fourth Illinois Infantry and Col. Jonas E. Thomas's First Tennessee Cavalry and a section of artillery—a 12-pounder field piece and 24-pounder howitzer—commanded by 1st Lt. Augustus Abel Gibson. Quitman's brigade consisted of the First Georgia and First Alabama regiments and the Baltimore infantry battalion of Maj. R. C. Buchanan. John R. Kenly, *Memoirs of a Maryland Volunteer: War with Mexico in the Years 1846–7–8* (Philadelphia: J. B. Lippincott, 1873), 210; G. W. Smith, *General Patterson's Route of March: Letter from the Secretary of War, Transmitting a Report on the Route of General Patterson's Division from Matamoras to Victoria,* 31st Cong., 2d sess., 1850, H. Exec. Doc. 13, 2.

George B. McClellan to Frederica M. English

ENGINEER CAMP NEAR MATAMORAS

20 DECEMBER 1846

My Dear Sister,

I received your kind letter a day or two ago. I have only time to scribble half a dozen lines to you today [to tell you] that we start from this place this afternoon or tomorrow morning, to march to Tampico, via Victoria.[1] I don't think that we will go to Tampico until the war is over. We will probably move on San Luis from Victoria. I hope so, for we will see a fight in that case. We leave behind us everything in the shape of heavy baggage such as trunks &c.—taking only a small carpet bag and a pair of saddle bags.[2]

As for being billeted on private families and all that sort of thing, I have been in a tent ever since I've been in the country, except about two weeks at Camargo where I had part of the palace of Don Jesus Guiste, said Hidalgo having departed for parts unknown, and said palace consisting of two remarkably uncomfortable rooms, utterly destitute of furniture, and if all the private families in Mexico are like those I have seen, I have merely to remark, "May the Lord deliver me from ever being billeted on a private family."

As for handsome Senoritas, the only passable one I've seen was in a rancho where the steamboat stopped for wood—and she was in the midst of dirt. Oh, no! It's very romantic perhaps at a distance, but there's a great deal of plain reality in it down here.

Well, I really must stop now. I received a day or two ago a letter from John dated October 30th, sent by a gentleman of Philadelphia. They were all well. You must not expect to hear from me during this march. Write when you can and direct to Tampico.

Believe me
your affectionate brother
Geo. B. McC
Love to Tom and kiss the "Nephew"

1. Patterson's division in fact did not depart Matamoras until 23 December 1846. Smith, *General Patterson's Route of March*, 2.

2. The complaints of the junior officers and enlisted men that they were not allowed ample transport for what they viewed as the necessities of the march were exacerbated by the lavish style in which senior officers, especially the generals, traveled. Pvt. George C. Furber noted that Patterson had three wagons for the use of himself and his staff and that Pillow had two. Each colonel had a wagon of his own, while a single wagon had to serve two companies of soldiers. One would "be surprised at the amount of kitchen furniture (enough for a good-sized hotel), bags, vegetables, champaign baskets, and cases of bottles, carpet bags, mattresses, bedding, trunks, etc." the generals carried on campaign. "It is a great thing to be an officer here," he marveled. George C. Furber, *The Twelve Months Volunteer; or Journal of a Private in the Tennessee Regiment of Cavalry* (Cincinnati: U. P. James, 1857), 322–33.

3

"A PERFECT DESERT FROM BEGINNING TO END"

VICTORIA TO TAMPICO

2 January 1847–5 March 1847

Following the strategically indecisive Battle of Monterrey, the Polk administration determined that further operations in northern Mexico were of no use in bringing the war to an end. Therefore it reluctantly authorized Maj. Gen. Winfield Scott to make a landing at Vera Cruz and from there to march to Mexico City. To raise an army, Scott drew heavily on Taylor's command, and the regiments selected to join the amphibious expedition were ordered to march from Matamoros to Scott's staging area at Tampico. The engineer company, as part of Maj. Gen. Robert Patterson's division, was assigned to the construction of a road passable for the column's artillery and baggage trains.

The labor was arduous. As Smith reported, "when the army would encamp, the pioneer party would go on until dark; then return to camp, and by daylight again be at the place we had left off." Nevertheless, one of the volunteers marveled at the "ease and quickness" with which the engineering company performed its function. "The most formidable obstacles were removed," he wrote, and the road was made practicable "by the united efforts of science, skill and labor."[1]

The sappers and miners followed the cavalry in order of march but preceded the infantry, artillery, and train so that when obstacles were encountered—"a huge rock in the road, an impassable ravine, or other impediment"—the company's bugler, who "could be heard a mile," sounded the halt for the entire column. A detail from the volunteer regiments, under the direction of the sappers, would then clear whatever blockage impeded the line of march.

Smith and McClellan "knew their business so well, and gave their directions so plainly, and they were followed with such quickness and union of effort," Pvt. George C. Furber of the First Tennessee Cavalry marveled, "that the obstacle was removed, the hill dug down, the ravine filled up, or the bridge strongly made, in less time than would, at home, be consumed in the preliminary consultation as to the manner of commencing the work.—It looked like magic; and it was the magical effect of knowledge, combined with connected labor."[2]

The engineers, reduced by disease to forty-five in number, arrived at Ciudad Victoria on 4 January 1847. On 12 January they were transferred to the division of regulars under Brig. Gen. David E. Twiggs and the following day began to make passable the road from Victoria south. The division arrived at Tampico on 23 January, having, according to Smith, "improved very decidedly" 364 miles of road and demonstrating that "in war hard work is sometimes necessary, and that fighting and marching are not all that are required of soldiers in the field."[3]

Of this march Colonel Totten reported, "the services of the company, constantly in advance and engaged in removing impediments and making the road practicable, were of great value." And although Smith recorded that to the volunteers of Patterson's division, the engineering company became known somewhat derisively as "the pick and shovel brigade," Furber referred to them as "a splendid company of regulars."[4]

1. Smith, *Patterson's Route of March,* 6; Furber, *Twelve Months Volunteer,* 297.

2. Furber, *Twelve Months Volunteer,* 298.

3. Smith, *Patterson's Route of March,* 2.

4. *Annual Report of the Chief Engineer to the Secretary of War,* 30th Cong., 1st sess., 1848, S. Exec. Doc. 4, 628; Smith, *Company "A,"* 5; Furber, *Twelve Months Volunteer,* 297.

Diary Entry

RANCHO PADILLO ON THE SOTO LA MARINA RIVER
2 JANUARY 1847

I "firmly believed" we would have a brush!—the devil I did!—and a pretty fool I was to think I'd have such good luck as that. I've given it up entirely. But I was right in the other—the whole column is Voluntario—and a pretty column it is too.

To go on with our affairs.—We reached Matamoras on the 8th [December] and encamped on the river [i.e., the Rio Grande] bank just below the Mexican batteries. Smith went down to the mouth again to select tools for the march, leaving me in command. After various orders and counter orders we were finally (December 21st) directed to appear upon the Plaza as early as possible in order to march to El Moquete, where General Pillow was encamped with the 3rd and 4th Illinois Volunteers. "Mind, Mr. Smith," said the old Mustang [General Patterson] the night before, "mind and appear as early as possible, so that you may not delay us"—all this with that air of dignity and importance so peculiarly characteristic of Mustangs; well we got up at daybreak and reached the Plaza a little after seven, immediately reported ourselves ready to start and were informed that we should wait for the guide who was momentarily expected. We were to march in advance, then the wagon train, then Gibson[1] with his artillery (a twelve-pounder field piece and twenty-four pounder howitzer) was to bring up the rear.

I waited and waited in the hot sun on the Plaza, watched the men gorging themselves with oranges, sausages etc., then took to swearing by way of consolation. When I had succeeded in working myself into a happy frame of mind (about one o'clock) old Abercrombie ordered Gibson to start in advance and our company to bring up the rear. I won't attempt to describe the beauties of forming a rearguard of a wagon train. Suffice it to say that the men straggled a great deal, some got rather drunk, all very tired. We reached the banks of El Arroyo Tigre about 8 o'clock (two hours after dark) and then encamped as we best could.[2]

I rode on in advance of the company to see El Tigre and found Gibson amusing himself by endeavoring to curse a team (a caisson) across the river, which (the caisson, not the river—well, both were, after all) had got mired in the middle. I rode back and met the company about one mile from the camp ground, struggling along—tired to death and straining their eyes to see water through the darkness. I consoled them somewhat by telling them that it was not more than a mile to the water, but they soon found that a mile on foot was a great deal longer than a mile on horse-back. However, we got there at last, pitched our camp, and soon forgot all our troubles in sound sleep.

I rode in advance next morning through the long wagon train to find a new ford. We crossed and encamped with General Pillow's Brigade.[3] Went down to Major Harris'[4] (4th Illinois) tent, where I had a fine drink of brandy and the unspeakable satisfaction of seeing a democratic Volunteer Captain (in his shirt sleeves) sit, with the greatest unconcern, on a tent peg for at least an hour. Gibson and I then went to Winship's tent where we found G. W. [Smith] and an invitation to dine with General Pillow.

During dinner it began to rain like bricks. We adjourned to Winship's tent, and the sight we presented would have amused an hermit. The water [was] about an inch deep in the tent, and we four sitting on the bed passing around a tumbler continually replenished from that old keg of commissary whiskey—oh lord! how it did fly 'round! and we were as happy a set of soldiers as ever lived "in spite of wind and weather."

"Whoa, Winship," says Gibson, "that's too strong," so he drank it all to keep us from being injured. Well, we amused ourselves in this way until dark—then we waded back to our respective domiciles (is a tent a domicile?) having previously seen old Patt make his grand entree in the midst of a hard rain—he in Dr. Wright's[5] covered wagon (looking for all the world like an old Quaker farmer going to market), his escort and staff dripping with the rain. We wondered why they looked so dismal and thought that it had not been such a horrid bad day after all!

This evening G. W. and myself had a grand cursing match over an order from the "stable" requiring a detail from our camp to pitch and unpitch the General's tents, etc. However, we sent them just about the meanest detail that they ever saw.

At this place our large army was divided into two columns. We moved at the head of the first column. General Pillow came on one day after us.

We started about 7:30 [23 December]—a bright sunny morning. Nothing of interest this day—the men improved in their marching. We encamped about three o'clock at Guijana, where there were two ponds of very good water. We had a beautiful spot for our encampment, and a most delightful moonlight evening. There is one house—hut rather—at this place.

From Matamoras to this place the road is excellent, requiring no repairs—chaparral generally thick on road side—one or two small prairies—road would be boggy in wet weather. From Matamoras to Moquete [is] about ten miles, from El Moquete to El Guijana about ten miles.

On the next day (December 24th) we marched to Santa Teresa, a distance of 27 miles.[6] It was on this march that we (i.e. Songo) made the "raise" on General Patterson's birds. He sent us four for supper. We ate as many as we could and had five left for breakfast—fully equal to the loaves and fishes, this.

We stopped for nearly an hour at Salina—a pond of rather bad water about half way to Santa Teresa—what a rush the Voluntarios made for the water![7] When we arrived we found the mustang crowd taking their lunch. As Songo had just then made one of his periodical disappearances we were left without any thing to eat for some time, but at last we descried him caracoling across the prairie on his graceful charger. The mustangs did not have the politeness to ask us to partake of their lunch, but when Songo did come our brandy was better than theirs anyhow.

At Santa Teresa the water was very bad—being obtained from a tancho. I bluffed off a volunteer regiment some 100 yards from our camp. As the Lieutenant Colonel[8] of this same regiment (3rd Illinois) was marching them along by the flank he gave the command "by file left march!"—to bring it on the color line. The leading file turned at about an angle of 80 degrees. "Halloa there" says the Colonel "you man there, you don't know how to file." "The h—l I don't" yells the man "d—n you, I've been marching all day, and I guess I'm tired."

Road good—passes principally through prairie—at Salina wood scarce in immediate vicinity of the water, plenty about three quarters of a mile from it. Wood not very plenty at Santa Teresa—enough however.

1. Augustus Abel Gibson, class of 1839, was assigned to Zachary Taylor's Army of Observation and took part in the battles of Palo Alto, Resaca de la Palma, and Monterrey. He

was prompted to first lieutenant in the Second Artillery on 18 June 1846 and served under Scott through the campaign against Mexico City.

2. This encampment was at Moquete, some eighteen miles south of Matamoras. Smith, *Patterson's Route of March*, 2.

3. This camp was at Guijana rancho, sixteen miles beyond Moquete. Ibid., 3.

4. Thomas Langrell Harris was an attorney in Petersburg, Illinois. He raised and was elected captain of Company F, Fourth Illinois Infantry and was elected major when the regiment was organized at Springfield.

5. Assistant Surgeon Joseph Jefferson Burr Wright received his medical degree at Jefferson Medical College, where he studied under McClellan's father. Early in the Civil War, Wright served as medical director on McClellan's staff during the West Virginia Campaign.

6. George C. Furber described Santa Teresa as "a pretty name, but a poor place as possible to be found.—Here were tough times for men and horses." Furber, *Twelve Months Volunteer*, 280.

7. Furber, one of the "Voluntarios," described the scene: "From dark until late at night, there was a continual crowd and quarrel for water, by the men;—and the horses bit and kicked one another in their efforts to get their heads into the narrow trough." Ibid., 284.

8. Lt. Col. Samuel D. Marshall of the Third Illinois Infantry was an attorney and the editor and publisher of the *Illinois Republican* in Shawneetown, Illinois.

Diary Entry

RIO LA CORONA

3 JANUARY 1847

We started before daylight and succeeded in getting clear of the volunteer camp by dint of great exertions. After marching about five miles through a fertile river bottom we reached the main branch of the Soto la Marina, a most beautiful stream of the clearest, coldest, most rapid water I ever saw—about sixty yards wide and three feet deep. Songo had some trouble in crossing without being washed off "Jim."

Padilla is situated on the banks of this stream—an old town rapidly going to ruin—with a quaint old Cathedral built probably 200 years ago, if not more. After marching about twelve miles more we reached the stream of La Corona, another branch of La Marina, similar in its character to the others. After working for about an hour on the banks we encamped on the further side. The Tennessee horse gave our men a "lift" over both the last streams—some of the Sappers had evidently never been mounted before.

Diary Entry

VICTORIA

4 JANUARY 1847

Very early we started for Victoria—and had to work our way through the camp of the Illinois regiments which was placed along the road. At last we cleared them and found ourselves marching by moonlight through a beautiful grove of pecan trees. I know nothing more pleasant than this moonlight marching, everything is so beautiful and quiet. Every few moments a breath of warm air would strike our faces—reminding us that we were almost beneath the Tropic.

After we had marched for about four hours we heard a little more yelling than usual among the Volunteers. Smith turned his horse to go and have it stopped when who should we see but the General and his staff in the midst of the yelling. We concluded that they must be yelling too, so we let them alone. This is but one instance of the many that occurred when these Mustang Generals were actually afraid to exert their authority upon the Volunteers.—Their popularity would be endangered. I have seen enough on this march to convince me that Volunteers and Volunteer Generals won't do. I have repeatedly seen a Second Lieutenant of the regular army exercise more authority over the Volunteers—officers and privates—than a Mustang General.[1]

The road this day was very good, and after a march of about seventeen miles we reached Victoria.[2] The Volunteers had out their flags, etc.—those that had uniforms put them on, especially the commandant of the advanced guard. Picks and shovels were put up—Generals halted and collected their staffs, and in they went in grand procession—evidently endeavoring to create the impression that they had marched in this way all the way—the few regular officers along laughing enough to kill themselves.

General Quitman[3] came out to meet General Patterson—but old Zach [Taylor], who arrived with his regulars about an hour before we did, stayed at home like a sensible man. We made fools of ourselves (not we either, for I was laughing like a wise man all the time) by riding through the streets to General Quitman's quarters where we had wine and fruit. Then we rode down to the camp ground—a miserable stony field—we in one corner of it, the "Continental Army" all over the rest of it. We at last got settled. About dark started over to General Taylor's camp. Before I had gone 200 yards I met the very person I was going to see—need not say how glad I was to meet him after a two months absence.

This reminds me that when at Matamoras—a day or two before we started on the march—we received the news of poor Norton's death.[4] I had written a letter to him the day before which was in my portfolio when I heard of his death. The noble fellow met his death on board the *Atlantic,* which was lost in Long Island Sound near New London on the 27th November 1846.[5] Captain Cullum and Lieutenant C. S. Stewart were both on board, and both escaped. Norton exerted himself to the last to save the helpless women and children around him—but in accordance with the strange presentiment that had been hanging over him for some time, he lost his own life. He was buried at West Point—which will seem to me a different place without him.

One night when at Victoria I was returning from General Taylor's camp and was halted about 150 yards from our Company by a Volunteer sentinel. As I had not the countersign I told him who I was. He said I should not go by him. I told him "Confound you, I won't stay out here all night." Said he, "You had no business to go out of camp." Said I, "Stop talking, you scoundrel, and call the Corporal of the Guard."—"I ain't got no orders to call for the Corporal and won't do it—you may, though, if you want." "What's the number of your post?" "Don't know." "Where's the Guard tent?" "Don't know."—As I was debating whether to make a rush for it, or to seek some softer hearted specimen of patriotism, another sentinel called out to me "Come this way, Sir!"—It appeared that the first fellow's post extended to one side of the road, and the last one's met it there.—"Come this way, Sir" said he, "Just pass around this bush and go in." "Hurrah for you," said I, "you're a trump, and that other fellow is a good for nothing blackguard."

1. Isaac Stevens echoed these sentiments in a letter dated 8 July 1847. In "difficult straits," he wrote to his wife, the council of subalterns from West Point was "sought and followed. The advice of lieutenants, even, is taken when that of general officers is disregarded." Stevens, *Life of Isaac Stevens,* 159.

2. G. W. Smith described Victoria as "a well built town of 2,000 or more inhabitants, situated on a small stream at the foot of the Sierra Madre." Smith, *Patterson's Route of March,* 4.

3. John Anthony Quitman, a plantation owner at Natchez, Mississippi, served in various state offices (including, briefly, that of acting governor) before volunteering for service in the War for Texas Independence in 1836. On 1 July 1846 President Polk appointed him a brigadier general of volunteers with command of a division in Scott's army. With the fall of Mexico City, he was appointed civil and military governor of the city. For his Mexican War service, Quitman received a brevet to major general and some support for the nomination to run for president on the Democratic ticket in 1848. Robert E. May, *John A. Quitman: Old South Crusader* (Baton Rouge: Louisiana State University Press, 1985).

4. Allen Higbee Norton, class of 1842, was a first lieutenant in the Fourth Infantry when he drowned during the sinking of the *Atlantic.*

5. The luxurious side-wheel steamer *Atlantic* sank on 27 November 1846, less than six months after being launched. Following a boiler explosion, the ship was driven aground on Fisher's Island, Long Island Sound, by a storm. Forty-five lives were lost.

Diary Entry

TAMPICO

13 JANUARY 1847

Left Victoria January 13th and arrived at Tampico on the 23rd. Wednesday January 13th. From Victoria to Santa Rosa four leagues. Road not very hilly, but had to be cut through thick brush; two very bad wet arroyos were bridged. Santa Rosa a miserable rancho—could only get a half dozen eggs and a little pig in the whole concern—good water in the stream.

Diary Entry

SANTA ROSA

14 JANUARY 1847

Started before daylight and before going 200 yards we landed in a lake—the road, or path, passed directly through it, and during the rest of the day it was necessary to cut the road through thick brush—no cart had ever been there before. Bridged two wet arroyos and encamped about sunset by a little stream.[1]

Just as enough water had been procured the stream was turned off—probably by the Mexicans. We had a stampede this day. Rode on about six miles with the guide. Country a perfect wilderness—not a rancho between Santa Rosa and Fordleone.

1. "The mule path was infamous," Smith complained. "No wagon had ever traveled that road—the rancheros have a tradition of a bull-cart that, it is said, once passed that way. I believe, however, that the story is not credited." Building materials too were sparse at best. According to Smith, these bridges were constructed of "ebony and mezquite [*sic*] wood—the material short, crooked, hard, and very heavy." Smith, *Company "A,"* 12; Smith, *General Patterson's Route of March,* 5.

Diary Entry

EL PASTOR

15 JANUARY 1847

Started early, road cut through a mesquite forest, many gullies, two bad arroyos before reaching El Pastor. Here General Twiggs[1] caught us, about 11 A.M., army encamped, but we went on.[2] I worked the road for about five miles, and started back at 4 [o'clock]. Smith and G** de L*** rode on about ten miles.[3] Road better, but very stony. "Couldn't come the cactus" over Guy de L*** this day. He (G. de L.) shot five partridges at a shot, which made us a fine supper.

1. David Emanuel Twiggs was, perhaps due to his favor with Andrew Jackson, appointed colonel of the Second Dragoons on 8 June 1836. This regiment served with Taylor's army at Palo Alto and Resaca de la Palma, earning its commander a brevet to brigadier general. At the Battle of Monterrey, Twiggs commanded one of Taylor's three divisions and served with distinction under Scott during the Mexico City Campaign, for which he was brevetted to major general.

2. Smith reported that "when the army would encamp, the pioneer party would go on until dark; then return to camp, and by daylight again be at the place we had left off."

3. William Seaton "Guy" Henry was graduated with the class of 1835 and assigned to the Third Infantry, rising to the rank of captain by 1846. He was brevetted to the rank of major for "gallant and meritorious conduct" during the Battle of Monterrey. During the war he wrote a series of descriptive letters that were published in the *Spirit of the Times,* a weekly newspaper published in New York, under the pseudonym G** De L***. His classmate George Gordon Meade described Henry as "a very good fellow," noting that "his recent productions since our march from Camargo have been quite spirited." These letters, revised and published in 1847 as *Campaign Sketches of the War with Mexico,* contain one of the best detailed accounts of the advance of Taylor's army from the Nueces to the Rio Grande. Meade, *Life and Letters,* 1:167–68.

Diary Entry

BANK OF THE RIO LA TULA

16 JANUARY 1847

Reveillé at 3—started at 4—arrived at end of preceding day's work just at daybreak. Road very stony in many places swore like a trooper all day—arrived at Arroyo Albaquila about 11 [A.M.]. Twiggs came up and helped us wonderfully by his swearing—got over in good time—cussed our way over another mile and a half—then encamped by the same stream—water very good.

Diary Entry

FORDLEONE

17 JANUARY 1847

Started before daybreak—road quite good—prairie land—arrived at Fordleone or Ferlón at about half after ten. Fine large stream of excellent water—good ford—gravelly bottom—gentle banks. 11 miles.

Diary Entry

EL PETRIL

18 JANUARY 1847

Reveillé at 3. Started long before daybreak—eyes almost whipped out of my head in the dark by the branches. Crossed the Rio Persas again at a quarter before seven—road rather stony in some places, but generally good. Great many palmetto trees—beautiful level country, covered with palmettos and cattle. "Struck" a bottle of aguardiente, or sugar cane rum. Made a fine lunch out of cold chicken and rum toddy—had another toddy when we arrived at our journey's end. Water from a stream, but bad. Porter sent us a peccary.[1] Rode on about three miles and found the road pretty good.

1. Fitz John Porter, class of 1845, was assigned to the Fourth Artillery and received brevets to captain and to major in Mexico for "meritorious service" at the battles of Molino

del Rey and Chapultepec. He was wounded at the Belen Gate, 13 September 1847. With the outbreak of the Civil War, Porter was commissioned a brigadier general of volunteers and assigned to command a division in the Army of the Potomac, becoming one of McClellan's most ardent and loyal supporters.

Diary Entry

EL PETRIL TO TAMPICO

19 JANUARY 1847–24 JANUARY 1847

On comparing notes at reveille, found that the rum and pelonay had made us all sick.[1] Started at 5 [19 January], road pretty good. Much open land, fine pasture—great deal of cattle. Reached Alamitos at about 9 A.M.—fine hacienda—good water, in a stream. Had a bottle of champagne for lunch—thanks to General Smith.[2]

From this place to Tampico, the principal labor consisted in making a practicable wagon road across the numerous arroyos—most of them dry at the time we passed: the banks very steep. Altamira is a pretty little town, one march from Tampico.[3] The road between them passes through a very magnificent forest of live oaks.

We encamped three miles from Tampico for about four days, and then moved into quarters in the town—the quarters so well known as "The Bullhead Tavarn."

1. Pelón Pelo Rico is a Mexican candy. McClellan's daughter, Mary, noted that her father's "sobriquet, in Mexico, among his intimate friends, was 'Pelon,'" which she took to mean "sugar." "On the march, when first arrived, he insisted upon eating a lot of the sugar arranged on even cobs and persuading his companions to eat it too. He was always fond of sweet things. They all became ill in consequence, and he more than any of them. After that they addressed him as 'Pelon' for he kept saying,—'Why it's Pelon, the *best* sugar—it can't hurt anyone.'" Vol. 108, McClellan Papers, Library of Congress.

2. Persifor Frazer Smith graduated from the College of New Jersey (now Princeton University) in 1815 and in 1819 moved to New Orleans, where he became Louisiana's adjutant general and commander of a battalion of militia. At the outbreak of the Mexican War, Smith was commissioned a colonel and given command of a brigade in Zachary Taylor's army. For his service at Monterrey, he was brevetted to brigadier general and given command of the elite new regiment of mounted rifles. Scott assigned him to the command of the First Brigade of Twiggs's division. For his contribution to the American victories at Contreras, Churubusco, Chapultepec, and the capture of the Belen Gate of Mexico City,

Smith was brevetted to major general. He was appointed to the armistice commission that arranged for suspension of hostilities and later served as military governor of the Mexican capital.

3. Twiggs's division bivouacked at Altamira on the night of 22 January, the tenth day of its march from Victoria. Smith described Altamira as "a large town [with a] fine old cathedral." Smith, *General Patterson's Route of March*, 6.

George B. McClellan to Dr. George McClellan

TAMPICO

25 JANUARY 1847

My dear Father,

We arrived here yesterday via Victoria—a march of about 400 miles. All well—men and all. I never was in better health in my life. I doubt whether I can get this off in time for the mail.

I write particularly to beg you to try and get me an appointment in one of the new Regiments of Infantry—a Captaincy I want. I would not like to accept anything less.[1]

Your affect. Son
Geo. B. McClellan

Crittenden sends his regards and is much obliged for your getting him [out here?]

1. On 11 February 1847, Congress created ten additional regiments of regulars to serve for the period of the war, the Ninth through the Sixteenth Infantry, the Third Dragoons, and the Regiment of Voltigeurs and Foot Riflemen.

Diary Entry

TAMPICO

24 JANUARY 1847–24 FEBRUARY 1847

Tampico is a delightful place—we passed a very pleasant time there and left it with regret.[1] We found the Artillery regiments encamped around the city. Many of the officers came out to meet us near Altamira. Champagne suppers were the order of the day (night I should say) for a long time. From Victoria to Tampico we were detached with Guy Henry's company of the 3rd—and Gantt's[2] of the 7th—Henry messed with us. When within about four day's march of Tampico we saw in front of us Mount Bernal, which is shaped like a splendid dome.

1. George Gordon Meade agreed. "Tampico is a delightful place, having fine cafes, and all the luxuries of a somewhat civilized town," he wrote, "much larger than I expected, and really quite delightful. There is a large foreign population of merchants, and in consequence the town has all such comforts as good restaurants, excellent shops, where everything can be purchased, and is in fact quite as much of a place as New Orleans." Meade, *Life and Letters*, 1:175, 177.

2. Levi Gantt, class of 1841, was a first lieutenant in the Seventh Infantry when he was killed in action at the Battle of Chapultepec.

George B. McClellan to Elizabeth Brinton McClellan

TAMPICO

4 AND 5 FEBRUARY 1847

Thus far are we, my dear Mother, on the road to the "long talked of City of the Montezumas," whither we are now in all probability to go by way of Vera Cruz. I've arrived here on the 24th of January, having been nearly a month and a week on the march from Matamoras. But one mail has been sent from here since our arrival, and that was the next day, so I had merely time to scrawl a line or two to Father, at the café.

As you must have received a short letter I wrote at San Fernando, I will merely say in reference to the march, that we got to Victoria on the 4th January, where we found General Taylor, who had arrived there about one hour before us. Our march to Victoria was principally over prairies and through chaparral, until we reached within a few days' march of Victoria, when the ground became more broken. It is a perfect desert from beginning to end. You meet occasionally with a village, such as San Fernando, Santander, &c., perfectly isolated, and that is all. We would sometimes encamp near a rancho, consisting generally of but one or two houses, but would march all day without seeing a sign of life. I cannot imagine how they subsist. Instead of finding cultivated lands around these inland towns, you find nothing but Chaparral. All is desert.

On Christmas day we had a very long and hot march. We encamped by a miserable little pond of wretched, muddy water and regaled ourselves on a tough beefsteak, mush, and hard bread for our Christmas dinner, instead of mince pie and roast turkey. However, we were very hungry and finished a bottle or so of Capt. Swift's good old Madeira, so we were not so badly off after all.

"Holloa, Smith," said I. "These confounded fat old cit[izen]s are stuffing themselves now with Turkies &c., but they can't enjoy them. Wait till next Christmas!"

New Year's Eve we spent near Santander and had a big eggnog!! We had one or two stampedes, but saw no enemy.

When we got to Victoria, we were attached to the 1st Division, Regulars, under Genl. Twiggs. Here we saw all the "heroes" and other big bugs, such as Old Zack, Maj. Mansfield, Bliss, Charley May, &c., &c.[1] You can form no idea of the pleasure it gave us to meet the Regulars after having been so long with the cursed volunteers. I found nearly all my own class here.

The Mexicans evacuated Victoria on the approach of our troops. I have been very much disappointed in not seeing Monterrey. Apart from the fact of its having been the scene of so bloody a conflict, it is said to be a most beautiful place, the country around, most delightful—cool water, fountains, orange groves &c., &c.

We had a great deal of hard work on the march from Victoria to this place, but arrived here safe and sound. We encamped about two and a half miles from the city, but in two or three days we (that is, our Company) came into town. We are now in quarters, and are very well situated. We live very well and are making ourselves quite comfortable. There is a theatre here, a splendid Café, a fine restaurant, and we meet (that is, the Regular officers) at the Café every evening and have fine times here.

I am tired of Tampico, for I like to be in motion. You have no idea of the charm and excitement of a march. I could live such a life for years without becoming tired of it. There is a great deal of hardship, but we have our own fun. If we have to get up and start long before daybreak, we make up for it when we gather around the campfires at night. You never saw such a merry set as we are. No care, no trouble, we criticize the Generals, laugh and swear at the mustangs and volunteers, smoke our cigars and drink our brandy—when we have any—go without when we have none.

A Regular officer has no habits—it is immaterial to him whether he gets up at 2 A.M. or 9—or whether he don't go to bed at all. When on a march we get up at 2 or 3, when we halt, we snooze it till 8 or 9. When we have cigars we smoke them, when we have none, we go without. When we have brandy, we drink it, when we have none, we make it up by laughing at our predicament. That is the way we have.

We are living now on the fat of the land—game, oysters, vegetables of all kinds, champagne, &c., &c., warm baths when we want them—in short, everything we want. Next week, we will probably be eating hard crackers and salt pork, drinking bad water, wearing the rowdiest kind of clothes on the que vive for a big fight at Vera Cruz.

The Mexicans are strengthening the place, additional troops are daily arriving there from the interior, but we will have no difficulty in reducing the place if they will ever commence the movement. We have about 7,000 men here. Worth's division has been at Brazos for about two weeks. The new regiments of Volunteers are being concentrated on an Island [Lobos] about 50 miles below here. General Scott is daily expected here from Brazos, and the affair appears to be rapidly approaching completion. I expect to be in the city of Mexico by the end of April or middle of May.

A shipload of volunteers were wrecked about thirty-five miles below here, three or four days ago. They were a portion of the Louisiana Volunteers.[2] A shipload of New York Volunteers are off the bar now. The Pennsylvania Regiment was at Brazos two weeks ago. I hope they do not put them in garrison on the Rio Grande. That would be too hard for the poor creatures. This place has been strongly fortified, so that it will not need a very large garrison to hold it. The old Mexican lines are very weak. They could not have stopped us a single day.

February 5th—A schooner arrived yesterday from New Orleans bringing seventy recruits for the 7th Infantry and one Company of the 1st and no mail!

General Smith told me last evening that a fight had occurred near Chihuahua between the Mexicans and a portion of General Harney's command. The Mexicans had intended to surprise the advanced guard of the detachment, but made a mistake and came in contact with the main body which knocked them soundly, taking all their artillery.

We know nothing whatever of General Scott's designs or when he will get here. He is expected every day and has been, ever since we arrived here. I am glad to see that another Regiment is coming from Pennsylvania.[3] I suppose Arthur feels like volunteering since he entertains such a mortal antipathy for the Mexicans!

I received a day or two before arriving here the letter sent by Crittenden, also those by General Scott. Crittenden's came via Monterrey. General Scott's overland from Matamoras to Victoria. Let me beg of you again not to send me any more letters by private conveyance. The mail is the only sure way. I would get letters by the mail, between two and three weeks after they left Philadelphia, while the very few I have received have taken about three months to reach me.

I am glad to hear that Father's neuralgia is better. How are the children? Arthur will be a man when I get back, I suppose. Tell him that I want to know whether he is too proud to wear frocks now.

I hope to spend next Christmas at home.

1. Joseph King Fenno Mansfield was graduated second in the class of 1822 and assigned to the engineers. He was brevetted to lieutenant colonel for his role in the Battle of Monterrey and to colonel for his service at Buena Vista. He was mortally wounded at the Battle of Sharpsburg, 17 September 1862, while commanding the XII Corps in McClellan's Army of the Potomac.

William Wallace Smith Bliss, known as "Perfect Bliss," was graduated with the class of 1833 and assigned to the Fourth Infantry. Winfield Scott, in his capacity as general in chief, appointed Bliss as Zachary Taylor's adjutant. In his own words, "knowing [Taylor] to be slow of thought, of hesitancy in speech, and unused to the pen," Scott "took care . . . to provide him, unsolicited, with a staff officer, [Bliss], his exact compliment, who superadded modest, quiet manners, which qualities could not fail to win the confidence of his peculiar commander, and on which usefulness entirely depended."

Charles Augustus May was commissioned into the Second Dragoons directly from civilian life. During the Battle of Resaca de la Palma, he led a daring if ill-considered charge against the center of the Mexican line that resulted—due to the lack of appropriate ammunition for the Mexican guns—in the overrunning of an enemy battery and the capture of a Mexican general. He received brevets through the rank of colonel for his service at Palo Alto, Resaca de la Palma, and Buena Vista but resigned from the army on 20 April 1861.

"Old Zack" is of course Zachary Taylor.

2. Col. Lewis G. DeRussy's regiment of Louisiana volunteers left Fort Jackson on 18 January 1847 for Lobos. On 2 February, however, a "tremendous norther" struck the transports, wrecking the *Ondiaka* some four miles south of Cape Roxza. The sloop of war *St.*

Mary's attempted a rescue, but Colonel DeRussy and his men, after being on shore for eight days and surrounded by Mexican forces, had marched for Tampico on 6 February, leaving tents and other camp equipage behind. Anonymous letter of a Louisiana volunteer, *New Orleans Commercial Times,* 22 Feb. 1847.

3. Two regiments of Pennsylvania volunteer infantry served with Scott's army: the First, which enlisted from December 1846 to August 1848, and the Second, which served from January 1847 to July 1848. Randy W. Hackenburg, *Pennsylvania in the War with Mexico: The Volunteer Regiments* (Shippensburg, Pa.: White Mane, 1992).

George B. McClellan to Dr. George McClellan[1]

TAMPICO

14 FEBRUARY 1847

We are still awaiting orders, my dear Father, in the same old place, knowing nothing more than we did when we arrived. General Scott is still at Brazos waiting for something or other—nobody can guess what. There are 7000 troops at this place, about 4000 more at Brazos, and a parcel of Volunteers at Lobos, but what in the world is to be done with us would be hard to tell. It is certainly too late to attack Vera Cruz, yet everything seems to point in that direction, so transports are said to be engaged for three months. Worth's division came from Saltillo to the mouth of the Rio Grande. The new regiments of Vols. are to concentrate at Lobos, an island about 70 miles below here.

For three weeks we have been here doing nothing, yet every day is of infinite importance to us. We cannot possibly land at Point Anton Lizardo in less than one month from this time. That would bring us to the middle of March, and the vomito commences its ravages at Vera Cruz early in April. The two or three weeks of the healthy season would be by no means sufficient to reduce the town and castle and to collect the wagons, teams, and stores necessary to carry our supplies on a march into the interior. I cannot believe that we will attempt Vera Cruz this spring.[2]

During the three weeks that we have spent here, the Mexicans have been strengthening the fortifications of the town, collecting supplies, and increasing the garrison. Three weeks ago there were but 1500 men in the town and castle, now there are 5000, and 11,000 more are daily expected, are probably there before this time. This is the fruit of our beautiful policy.

It would have been just as easy to have moved upon this place in the month of October as in that of January. It would have been just as easy to have called out the whole of 50,000 men, authorized by Congress, in the month of July, and to have encamped and drilled them in some healthy place (at the mouth of the Ohio, for instance) as to have called out [illegible] Regiments in the month of December.

What would have been the result? We would have attacked Vera Cruz in the latter part of November with about 8000 Regulars and some 40,000 Volunteers in a tolerable state of discipline. Long before this, or the latter part of January at furthest, we would have been in the city of Mexico, or else a peace would have been made on the most favorable terms.

How are we now situated? There are about 3000 Regulars here, as many more at Brazos, about 4000 Volunteers here, whose term of service expires on the 1st June. There will be between 5 and 6000 more Volunteers at Lobos, undrilled, undisciplined, and a portion unarmed—more than all, utterly unacclimated—the sickly season is rapidly approaching, the wet season will shortly set in, and that will render it utterly impossible to march from here to Monterrey, Victoria, or Matamoras. The road from here to the city of Mexico, and that to San Luis, are utterly impracticable for wagons and artillery—to move without them would be utter madness! Can you imagine a more disagreeable situation? Can you imagine a weaker head than that which has brought us into this predicament?

I can think of but one solution of the enigma—we must have peace. Negotiations must soon be pending; if it is not so, I can only hope that heaven may save the credit of our country's arms—nothing else will.

The war has not commenced—if we have peace now, I shall always maintain that we have had no war; we had two brushes with the enemy, an erratic movement, ending in a butting of heads against a stone wall, then a grand rush for the seaboard. The operations which have taken place—if worthy of the name—are disgraceful to us, the weakest government of the old world, the pettiest Italian town, would have displayed more energy—the merest tyro in the Military Art would have done more than we have done—but what can we expect of a government afraid for its popularity, striving to withdraw its diminished head from the messes which envelope it, too mean and foolish to spend a penny, when that economy involves the expenditure of thousands in the end; of a government which regards a military education as an indication of unfitness for military command; where new regiments are officered by deficients of the Military Academy, friends of politicians and barroom blackguards, to the exclusion of those officers who have served in the field and earned their title to promotion in actual campaign; with whom old age and political favor are regarded as the requirements of a general? Who can account for the strange, glaring inconsistency of which the whole nation is guilty?

We hear resounding from one end of the Union to the other the cry—The Volunteers are the right arm of the nation, to them will we commit our honor and our fame! Every one believes, or what is practically the same thing, pretends to believe, that although a long course of study is necessary to fit a man for the pursuit of any civil profession, yet any one can become a good officer by putting on a military coat—that the change of dress can produce as wonderful an alteration in the character of the man more wonderful than that of the colors of the chameleon. The climate of Spitzbergen and Arabia are not more different than the characteristics of a Civil and a Military man, and as well might we expect that climates to change in an hour as to see a citizen become a good officer without years of training.

Our dear sovereigns, and those who lead them by the nose, would indulge in the pleasant delusion that any one who is elected as officer of Volunteers becomes just as good an officer as those who have for four years gone through the strict discipline of West Point. They do not confine themselves to subaltern officers, but give us generals taken from behind the counter and the desk.

Yet propose to any of these men to give the control of our armies to some one of our younger officers—e.g., a Captain or Lieutenant who has not only received a thorough Military Education at our National Academy, but has also served in the field, and you are at once assailed with the cry, "He is too young, he has no experience!"

Shame, shame upon those who would delude themselves or others with such a lie! Who but the most egregious fool would believe, who but the most arrant knave would say that years spent in shaving notes, saving dollars, or poring over the thick volumes of law give a man that kind of experience which is necessary for a general? [Phrase illegible] tells us—

and history confirms the assertion—that he who must wait until time has withered his locks before he is fit to assume command can never be competent, not even should he live to the age of Methuselah. Nature and Education must do much, and these must be learned from experience in the field.

1. Dr. George McClellan was born on 22 December 1796 in Woodstock, Connecticut. After receiving his M.D. degree from the University of Pennsylvania in 1819, he went into private practice in Philadelphia and established the nation's first free eye clinic, the Institution for the Diseases of the Eye and Ear. In 1824 he proposed the revolutionary idea of teaching medical students by having them observe experienced doctors treating patients and of supervised student participation in the care of patients. Doctor McClellan thus became the founder of Jefferson Medical College and reshaped the way medicine is taught throughout the world. He held the chair of surgery at Jefferson until 1838, when a conflict with the Board of Trustees led to his dismissal from the faculty. McClellan responded by establishing in 1839 the Medical Department of Pennsylvania College. In 1843, however, financial and other difficulties compelled the entire faculty to resign. J. Thomas Scharf and Thompson Westcott, *History of Philadelphia, 1609–1884*, vol. 2 (Philadelphia: L. H. Everts, 1884), 1645; McClellan Family Papers, Thomas Jefferson University Archives.

2. McClellan of course was wrong. From Scott's landing below Vera Cruz on 9 March until the city fell on 28 March, only nineteen days passed. Indeed the whole tenor of McClellan's complaint seems to foreshadow his exaggerations of enemy strength and his ponderous and ultimately fruitless advance on Richmond in the spring of 1862.

George B. McClellan to Dr. George McClellan

TAMPICO

22–23 FEBRUARY 1847

My dear Father,

We are hourly expecting to embark for Vera Cruz, first concentrating at Lobos. A Norther, which blew yesterday prevents the embarkation. General Scott arrived here on Friday morning and left on Saturday. He looked very well, indeed. Enquired particularly if I had heard from you and desired to be remembered to you when I wrote next.

I dined with General Patterson yesterday, who particularly desired to be remembered to you. He tries to be very kind and polite to me. The old

General (Scott) was very kind in the few moments I saw him. Colonel Totten was there with him. Captain Swift has not joined us, but we expect to meet him at Lobos. Colonel Totten told me that the Captain was not well enough for active service, but that he had not the heart to order him home before Vera Cruz was over.

The expedition is more fully fitted out than anything has ever been in this country. A miracle alone can save the Castle. The Army is in splendid spirits, and if the "Old Army of the Rhine" (Twiggs's division) don't astonish the Mexicans, I am mistaken![1]

Our Company is in fine health and spirits—I know it will do its duty, notwithstanding all those letters written by one or two of the rascals who have shirked and got off to the States, and I know that the men will stand by their officers to the last. We will probably have a pretty affair at Vera Cruz.

As a matter of course there are every day a dozen contradictory rumors about the evacuation of Vera Cruz, reinforcements being received, &c. &c. We pay no attention to them here. This is probably the last letter you will receive until the siege is over, which will probably be the case before you get this. Until then,
believe me to be your
Affectionate son
Geo. B. McClellan

P.S. Smith sends his regards. Give my love to Mother, John, and the babies. I am glad to hear that you have really begun the Surgery. It is a duty you owe to your family, yourself, and your profession. I hope to hear that it is finished when I return home.

Feb 23rd—We sail tomorrow for Lobos.

1. On 12 October 1808 Napoleon instituted the Army of the Rhine, made up for the most part of new conscripts, most of whom were no more than sixteen years of age. Officers for this stopgap force were supplied by stripping support facilities of their officers, recommissioning pensioned officers, commissioning the cadets of the various military academies before their graduation dates, and commissioning numerous sergeants and warrant officers. David Chandler, *The Campaigns of Napoleon* (New York: Macmillan, 1966), 334, 668–70.

George B. McClellan to Frederica M. English

TAMPICO

22–23 FEBRUARY 1847

My dear Sister,

The Army is on the point of embarkation for Vera Cruz. We are only awaiting until the effects of yesterday's Norther have subsided to commence the movement. The first division, regulars, move first, then our Company, then the Artillery. The Baltimore battalion and the Louisiana Volunteers are left in garrison at this place. The Lord deliver them, poor fellows.

The expedition is most splendidly equipped, far better than anything of the kind ever has been in this country. The Army is in splendid health and spirits and most anxious for a big fight.

23rd We sail for Lobos tomorrow and as I am just as busy as a man can be in preparing to "vamoose" this ranch. Captain Swift will join us at Lobos. Give my love to Tom, kiss the baby for me, and believe me to be
Your affectionate brother
Geo B. McClellan

I'll write when the fight is over and let you know what kind of an affair it is.

Diary Entry

LOBOS

24 FEBRUARY 1847–2 MARCH 1847

We left Tampico at daylight on the 24th [of] February on board a little schooner called the *Orator*—a fast sailer, but with very inferior accommodations.[1] I really felt sorry to leave the old "Bullhead Tavarn" where I had passed so many pleasant moments. The view of the fine city of Tampico as we sailed down the river was beautiful. Its delightful rides, its beautiful rivers, its lagoons and pleasant Café will ever be present to my mind. Some of the happiest hours of my life were passed in this same city—Santa Anna de Tamaulipas.

On arriving at Lobos we found that we had arrived a day in advance of the "Army of the Rhine," which had started a day before us. Lobos is a small island formed by a coral reef—about eighteen or twenty miles from the shore, forming under its lee a safe but not very pleasant anchorage. I went on shore but found nothing remarkable. Some sixty vessels were there when we started.

1. The schooner *Orator* was chartered at Tampico to transport troops to and from that port. *Report of the Secretary of War,* 30th Cong., 1st sess., 1848, S. Exec. Doc. 59, 44.

Diary Entry

ANTON LIZARDO

3–5 MARCH 1847

At last the order was given to sail for Point Anton Lizardo. We sailed next day [3 March] but one after the generals and arrived before any of them except Twiggs. We ran on the reef under the lee of Salmadina Island, were immediately taken off by the navy boats which put us on shore where we were very kindly received by the Rocketeers.[1] It was a great relief to get rid of that confounded red and white flag—"send a boat with an officer"—and the disagreeable duty of reporting to the 'Generál en Géfe' every morning.

A French sailor of the *Orator* undertook to pilot us and carried us on a reef of what he called Sacrificios but what turned out to be Anton Lizardo.[2]

1. The Ordnance Department adopted a military rocket, developed by American inventor William Hale, and employed a battery of these weapons at the siege of Vera Cruz under the command of Maj. George Henry Talcott. Although the rockets created something of a spectacle, their usefulness was slight. Lester R. Dillon Jr., *American Artillery in the Mexican War, 1846–1847* (Austin, Tex.: Presidial, 1975), 17, 39.

2. The *Orator,* according to Smith, was commanded by "a brave little Frenchman" who had little acquaintance with the gulf coast of Mexico and who lacked the navigational instruments necessary to make a proper landfall.

4

"WE HAD DONE MORE THAN ALL THE REST"

VERA CRUZ

9 March 1847–2 April 1847

When Scott's army splashed ashore below Vera Cruz on 8 March, the engineer company, attached to a regiment of regulars commanded by Brig. Gen. William J. Worth, was among the first to land. Captain Swift, who had rejoined the company at Anton Lizardo, was "too feeble to walk across the cabin of his vessel without assistance," and McClellan had advised him that he ought not to attempt landing with the company. As McClellan told Smith, however, "he looks upon me as a boy, and I have no influence with him in this matter." McClellan, therefore, urged Smith to discourage the captain from personally leading the assault. When Smith demurred, fearing that urging the captain to stand down would be interpreted as a dishonorable attempt to gain command of the company, McClellan persisted, adamant that "this case is beyond mere delicacy" and insisting that "leading the company ashore will kill him."[1]

Smith did attempt to dissuade his captain from what he was sure was a suicidal act of heroism, but Swift insisted, leading twenty men of the company into a surfboat and, with McClellan commanding the remainder of the engineers in another boat, rowing with them ashore. The effort, as McClellan had predicted, exhausted his remaining strength. Swift, Totten reported, "strove nobly, but vainly, against an inexorable disease. A too ardent sun prostrated him at once, depriving the country of his services at a moment when his high and peculiar talents would have been of the greatest value." On the day following the landing, Swift was returned to Scott's flagship, *Massachusetts,* and from there sailed to New Orleans, where he died on 24 April 1847.[2]

Command of the engineer company, then "in a state of great despondency and discontent," devolved upon G. W. Smith, who according to Isaac Ingalls Stevens, demonstrated "great power of command," and "by his judicious management he breathed into [the sappers and miners] the breath of life, raised the spirits of the men, and inspired them with hope and confidence."[3]

Its first duty under Smith, undertaken on 11–12 March, was to cut a road through the chaparral north from the army's landing place on Collado Beach to Vera Cruz and to interdict the aqueduct that supplied fresh water to the walled city. Thereafter the company was constantly employed in the construction of siege lines and the laying of heavy batteries with which to reduce the Mexican stronghold. As Colonel [Ethan Allen] Hitchcock observed, the approach to the city was "conducted under the direction of scientific Engineers and everything has proceeded according to known rules of the Art of War." But a strong sense of urgency impelled Scott's army, for the dreaded yellow-fever season was only a few weeks away, and it was imperative that the Americans capture the city and its castle and move onto higher ground inland before the onset of the deadly *vomito.*[4]

By 13 March, as one volunteer recorded in his journal, "the sappers and miners [were] hard at work ditching, to get a favorable position for a battery." In fact the sappers and miners played a major role in this vital siege, and all of the men, Stevens wrote, "exhibited an extraordinary gallantry, and were all placed in the position of non-commissioned officers. Each man had direction of a working party, and in the execution of that duty they retained their arms and gave directions to the men."[5]

From the moment the company landed, Totten reported, it "was engaged in the most severe and trying duties, in opening paths and roads to facilitate the investment, in covering reconnoissances, and in the unceasing toil and hardship of the trenches. The total of the company was so small, and the demands for its aid so incessant, that every man may be said to have been constantly on duty, with scarcely a moment for rest or refreshment."[6]

On 16 March McClellan, Smith, and 2d Lt. John Gray Foster, who had been attached to the company at Anton Lizardo, each took one-third of the engineer company to reconnoiter the best position on which to plant the army's heavy batteries to begin the shelling of Vera Cruz. Their exposed position, however, made the engineers susceptible to Mexican artillery fire. Little daunted, McClellan returned to camp that night with his clothes "very much torn," though according to Hitchcock he merely laughed and said that "the Mexicans had been firing at his party nearly all day without hitting a man."[7]

Having selected a location to "the right of the city and within about 450 yards of its works," the company of engineers commenced to mount the guns the following morning, and on the night of the 18 March, the men began to excavate trenches and earthworks. This labor was conducted under the direct fire of twenty-two pieces of Mexican artillery "placed in a strong entrenched camp to oppose our operations, and surrounded by every advantage of ground, besides immense bodies of cavalry and infantry." Smith recorded in his diary on 19 March that McClellan "had been for some hours detached" when he "heard just ahead the sharp firing of musketry; and immediately after met Captain [John] McClellan, of the topographical engineers, and Lieutenant [George B.] McClellan, of the engineer company, returning on horseback." Having come suddenly upon a strong enemy picket, the two had been fired upon, with Lieutenant McClellan having his horse shot from under him.[8]

The engineers' labor was further exacerbated by what Stevens called "continued and distressing northers," which destroyed works they had just constructed, and by the unpleasant living conditions that the whole army endured among the sand dunes. The troops, wrote McClellan's West Point classmate Dabney H. Maury, were "nearly three weeks in the sand hills without change of raiment," and their "opportunities for bathing were very limited." Maury was especially exercised by the fleas, declaring that "if one were to stand ten minutes in the sand, the fleas would fall upon him in hundreds." In fact, he recalled, "Smith, and McClellan, slept in canvas bags drawn tight about their necks, having previously greased themselves all over with salt pork. Perhaps the fleas did not partake of them, but they made up for it by regaling themselves of us of the line who had no canvas bags."[9]

Despite such hardships, by 22 March Scott was ready to demand the city's capitulation or to destroy its walls with heavy artillery. At 8:00 that morning Totten sent McClellan to the trenches to assist in the bombardment. By 2:00 P.M. the army's mortars and a battery of naval ordnance were trained on the city's defenses and ready to fire.[10]

Capt. Joseph E. Johnston of the topographical engineers carried a flag of truce to the garrison, demanding its surrender. Gen. of Brigade Juan Morales, the commander of the Mexican army in Vera Cruz, disdained Scott's terms, and Smith recorded in his diary that "the flag had hardly commenced its return from the town when a few spiteful shots from Santiago at my party on the magazine told us plainly enough what the reply had been."[11]

Within half an hour, therefore, the heavy American mortars and naval guns began to knock the city's walls into rubble. For two days the garrison and citizens of Vera Cruz withstood the terrible fire, while Scott's lines, laid out and constructed by his engineers, crept closer. The end came later that day. "When the white flag was shown at Vera Cruz we were overjoyed and greatly comforted," Maury recalled.[12]

Totten was rightfully proud of his engineers' contribution to the American cause during the siege. All, he reported to Secretary Marcy, "behaved so well and did their duty so faithfully and zealously that it would be invidious to distinguish between them," but he did have particular praise for the engineering company. In his report to Scott, the chief engineer wrote of "the highly meritorious deportment and valuable services" of the sappers and miners. "Strenuous as were their exertions, their number proved to be too few, in comparison with our need of such aid. Had their number been fourfold greater, there is no doubt the labors of the army would have been materially lessened and the result expedited."[13]

Moreover, notwithstanding he had stated that "it would be invidious to distinguish" among his officers, Totten told the secretary of war that he had witnessed "the great exertions and services of this company, animated by, and emulating the zeal and devotion of its excellent officers, Lieutenants Smith, McClellan, and Foster."[14]

Even Beauregard, ever jealous of professional acclaim and personal honor, jotted in his diary that during the siege the engineering officers had "displayed constantly that activity, intelligence and gallantry for which they are so well renowned (conspicuous among them, however, were Captain Lee and Lieutenant McClellan)."[15]

Scott too had high praise for his engineers. As McClellan returned to the camp of the sappers and miners on the morning of 25 March—shortly before the capitulation of Vera Cruz—he stopped at Colonel Totten's headquarters to inform his commander of conditions at the front, and Totten in turn directed the young engineer to make the same report to General Scott. "I found him writing a dispatch," McClellan recalled. "He seemed to be very much delighted and showed me the last words he had written which were 'indefatigable Engineers.'"

The enlisted men of the engineer company also came in for commendation. To his father a few days after the capitulation of the city, Stevens boasted that the rank and file of the company "are the pride of the whole army, confessedly the best soldiers in the army."[16] Even the despised volunteers lauded the efforts of the sappers and miners. "Great praise is due to Gen'l Scott & the Artillery & Engineer Corps," wrote Capt. Sydenham Moore of the First Alabama Volunteers, "for their conduct here and the result is a glorious victory for our arms."[17]

Nevertheless the engineers felt that the company had not received the credit it was due. Stevens, Totten's adjutant, recorded in his diary that when the army entered Vera Cruz, "the engineer company, although it had preeminently distinguished itself for gallantry and general conduct throughout the whole operation of the investment and siege, had no place assigned to it in the ceremonies of either the surrender or the entrance."[18]

And despite all of the acclaim heaped upon his company in general and upon himself in particular, McClellan's exaggerated sense of honor was piqued when, following the Mexican surrender, the commanding general and the chief engineer failed to continue to sing his praises. During the siege "we were needed and remembered," he wrote in a characteristic fit of overweening ego, "the instant the pressing necessity passed away we were forgotten. The echo of the last hostile gun at Vera Cruz had not died away before it was forgotten by the Commander in Chief that such a thing existed as an Engineer Company."

1. Smith, *Company "A,"* 14.

2. *Annual Report of the Chief Engineer to the Secretary of War,* 30th Cong., 1st sess., 1848, S. Exec. Doc. 4, 628.

3. Stevens, *Life of Isaac Stevens,* 216–17.

4. Ethan Allen Hitchcock to Elizabeth Nichols, 27 Mar. 1847, Ethan Allen Hitchcock Papers, Manuscript Division, Library of Congress; Smith, *Company "A,"* 195; Adrian George Traas, *From the Golden Gate to Mexico City: The U.S. Army Topographical Engineers in the Mexican War, 1846–1848* (Washington, D.C.: Center of Military History, 1993), 184.

5. Steven R. Butler, ed., *The Eutaw Rangers in the War with Mexico: The Mexican War Journal and Letters of Capt. Sydenham Moore and the Mexican War Journal of Pvt. Stephen F. Nunnalee, Company D, First Regiment of Alabama Volunteers* (Richardson, Tex.: Descendants of Mexican War Veterans, 1998), 105; Isaac Ingalls Stevens to Father, 11 Apr. 1847, quoted in Stevens, *Life of Isaac Stevens,* 118.

6. Totten to Marcy, "Report of the Chief Engineer," in *Report of the Secretary of War,* 31st Cong., 1st sess., 1849, H. Exec. Doc. 1, 279–80.

7. Ethan Allen Hitchcock, *Fifty Years in Camp and Field: Diary of Major-General Ethan Allen Hitchcock, U.S.A.,* ed. W[illiam] A[ugustus] Croffut (New York: G. P. Putnam's Sons, 1909), 241.

8. Smith, *Company "A,"* 23, 40–41.

9. Maury, *Recollections of a Virginian,* 34.

10. Stevens, *Campaigns of the Rio Grande,* 53.

11. William Starr Myers, ed., *The Mexican War Diary of General George B. McClellan* (Princeton: Princeton University Press, 1917), 66–67.

12. Hitchcock, *Fifty Years in Camp and Field,* 244; Stevens, *Life of Isaac Stevens,* 113; Maury, *Recollections of a Virginian,* 34; Bradley T. Johnson, ed., *A Memoir of the Life and Public Service of Joseph E. Johnston* (Baltimore: R. H. Woodward, 1891), 291.

13. Joseph G. Totten to Winfield Scott, 28 Mar. 1847, in *Report of the Secretary of War,* 30th Cong., 1st sess., 1847, H. Exec. Doc. 8, 245. See also *Annual Report of the Chief Engineer to the Secretary of War,* 30th Cong., 2d sess., 1848, H. Exec. Doc. 1, 278–81.

14. Totten quoted in Hillard, *Life and Campaigns of George B. McClellan,* 21; Isaac Ingalls Stevens to Wife, 3 Apr. 1847, quoted in Stevens, *Life of Isaac Stevens,* 116.

15. Beauregard, *With Beauregard in Mexico,* 30–31.

16. Isaac Ingalls Stevens to Father, 11 Apr. 1847, quoted in Stevens, *Life of Isaac Stevens,* 118.

17. Butler, *Eutaw Rangers,* 65.

18. Isaac Ingalls Stevens Diary, 29 Mar. 1847, quoted in Stevens, *Life of Isaac Stevens,* 118.

Diary Entry

VERA CRUZ

9–14 MARCH 1847

On the morning of the 9th of March we were removed from the *Orator* to the steamer *Edith,*[1] and after three or four hours spent in transferring the troops to the vessels of war and steamers, we got under weigh and sailed for Sacrificios.[2] At half past one we were in full view of the town[3] and castle,[4] with which we soon were to be very intimately acquainted.

Shortly after anchoring the preparations for landing commenced, and the 1st (Worth's) Brigade[5] was formed in tow of the *Princeton*[6] in two long lines of surf boats[7]—bayonets fixed and colors flying. At last all was ready, but just before the order was given to cast off a shot whistled over our heads. "Here it comes" thought everybody, "now we will catch it."

When the order was given the boats cast off and forming in three parallel lines pulled for the shore, not a word was said—everyone expected to hear and feel their batteries open every instant. Still we pulled on and on—until at last when the first boats struck the shore those behind, in the fleet, raised that same cheer which has echoed on all our battlefields—we took it up and such cheering I never expect to hear again—except on the field of battle.

Without waiting for the boats to strike, the men jumped in up to their middles in the water and the battalions formed on their colors in an instant—our company was the right of the reserve under Colonel Belton.[8] Our company and the 3rd Artillery ascended the sand hills and saw—nothing.

We slept in the sand—wet to the middle. In the middle of the night we were awakened by musketry—a skirmish between some pickets. The next morning we were sent to unload and reload the "red iron boat"—after which we resumed our position and took our place in the line of investment.[9] Before we commenced the investment, the whole army was drawn up on the beach. We took up our position on a line of sand hills about two miles from the town. The Mexicans amused themselves by firing shot and shells at us—all of which (with one exception) fell short.

The sun was most intensely hot, and there was not a particle of vegetation on the sand hills which we occupied. Captain Swift found himself unable to stand it, and at about half past twelve gave up the command to G. W. Smith and went on board the *Massachusetts*[10] that same afternoon. He did not resume the command, but returned to the United States. He died in New Orleans on the 24th of April.

About one, we were ordered to open a road to Malibran (a ruined monastery at the head of the lagoon). The Mohawks had been skirmishing around there, but, as I was afterward informed by some of their officers, that they fired more on each other than on the Mexicans.

After cutting the road to Malibran we continued it as far as the railroad—a party of Volunteers doing the work and some twenty-five of our men acting as a guard. When we arrived at the railroad, we found it and the chaparral occupied by the Mexicans. Our men had a skirmish with them—charged the chaparral and drove them out of it.

We returned to Malibran and bivouacked on the wet grass without fires—hardly anything to eat—wet and cold. Got up in the morning and resumed our work on the road—from the railroad to the "high bare sand hill"—occupied by the Pennsylvanians the night before. The work was very tedious, tiresome and difficult—the hill very high and steep—and the work not at all facilitated by the shells and shot that continually fell all around us. At last we cut our way to the summit tired to death.

A.M.—rifleman was killed this morning by a 24-pound shot—on top of the hill. Lieutenant Colonel Dickenson[11] and some few Volunteers were wounded by escopette balls.[12]

I was sent up in the morning to find the best path for our road and just as I got up to the top of the hill the bullets commenced whistling like hail around me. Some Lancers were firing at the Volunteers—who were very much confused and did not behave well. Taylor's Battery[13] and the rest of Twiggs's Division moved over the hill towards their position on the left of the line. Worth's Division (or Brigade as it was then called) occupied the right of the investment, the Mohawks under Patterson the centre, and Twiggs the left. After resting our men at Malibran, we moved back to our old position with the 3rd Artillery, where we bivouacked.

I had observed on the preceding day the position of the aqueduct supplying the city with water. I told Lieutenant Beauregard[14] next morning what I had seen. He reported it to Colonel Totten[15] and Smith and myself were ordered to cut off the water, Foster[16] remaining at home. We took a party, cut off the water, Smith exploded a humbug of Gid Pillow's, and we started on a reconnoitering expedition of our own.

I stopped to kill a "slow deer" [i.e., a cow] and Smith went on. I then followed him with three men and overtook him a little this side of the cemetery. We went on to within 900 yards of the city and at least a mile and a half in advance of the line of investment—ascertained the general formation of the ground and where to reconnoiter. We returned after dark, Foster much troubled as to what had become of us. It was upon reporting to Colonel Totten on this night (12th) that he said that I and

G. W. were the only officers who had as yet given him any information of value—that we had done more than all the rest, etc., etc. All forgotten with the words as they left his mouth—vide his official report of the siege.[17] G. W. and myself will never forget how we passed this blessed night—(new fashioned dance).

On the next day Foster was sent after our baggage and camp equipage. I was ordered to move the company and pitch the tents on a spot on the extreme right. [G. W.] Smith went out with Major Smith[18] to where we had been the night before, but went no further toward the city than we had been.[19]

1. The screw steamer *Edith* is said to have been the first propeller-driven craft built for the U.S. government, serving as a transport in the Gulf of Mexico during the war with Mexico.

2. William Augustine, a first lieutenant in the Third Artillery, described the island of Sacrificios as "a little sand bank a mile from shore on the left of the harbor as you enter, affording a refitting place & a good anchorage in the event of storms." James M. McCaffrey, ed., "America's First D-Day: The Vera Cruz Landing of 1847," *Military History of the West* 25, no. 1 (Spring 1995): 54.

3. Vera Cruz, Mexico's most important port and the gateway to Mexico City, was an obvious strategic prize. In November 1846 Winfield Scott presented to the Polk administration a proposal to take the city by siege, following an amphibious landing to the south, out of range of the heavy guns of Vera Cruz's guardian fortress, the castle of San Juan de Ulúa. With Polk's approval, Scott moved an army of 15,000 men by sea from Tampico, landing on Collado Beach below the fortress city on 9 March 1847.

4. San Juan de Ulúa, some 500 yards offshore from Vera Cruz, was first utilized in the defense of Mexico's principal port in 1519, when Hernán Cortéz established a small fort there. In 1771 the Spanish viceroyalty began construction on the coral reef of a massive stone castle, capable of housing a 2,500-man garrison and mounting 130 heavy-caliber guns. With the fear of an impending invasion by U.S. forces, Mexican authorities ordered 1,200 infantrymen as well as some 450 artillerists into the castle under the command of Gen. of Brigade José Durán. Considering his fortifications invulnerable to naval attack and believing Vera Cruz impregnable, the commandant of the castle neglected to lay in a supply of fresh water and provisions at the beginning of the siege. Thus the capture of Vera Cruz placed the castle entirely at Scott's mercy, and it was surrendered with the city on 27 March 1847.

5. William Jenkins Worth—"Bold Billy Jenkins" to McClellan—was commissioned a first lieutenant in the Twenty-third Infantry at the beginning of the War of 1812 and, despite a limited education, was appointed an aide de camp to Brig. Gen. Winfield Scott. In 1820 he was appointed commandant of cadets at West Point, serving in that capacity until 1828. At the outbreak of the war with Mexico, Worth was second in command of Zachary Taylor's Army of Occupation along the Rio Grande. His division was primarily responsible for the

U.S. victory at Monterrey, for which he was brevetted to major general. He commanded a division of Scott's army with great distinction during the Mexico City Campaign. Edward S. Wallace, *General William Jenkins Worth* (Dallas: Southern Methodist University Press, 1953).

6. The USS *Princeton,* a 1,046-ton screw steamer, was designed by the famed naval architect John Ericsson and constructed at the Philadelphia Navy Yard. It was commissioned in September 1843. On 28 February 1844, while demonstrating her heavy-shell guns near Washington, D.C., a cannon exploded, killing Secretary of State Abel P. Upshur, Secretary of the Navy Thomas Gilmer, and several other distinguished visitors. *Princeton* was assigned to the Gulf Squadron, serving in the war with Mexico until 1847.

7. For his amphibious assault below Vera Cruz, Scott ordered 141 landing craft. Designed by Lt. George M. Totten, USN, these double-ended, broad-beamed, flat-bottomed surfboats were built in three sizes so that they could be nested for transport. K. Jack Bauer, *Surfboats and Horse Marines: U.S. Naval Operations in the Mexican War, 1846–48* (Annapolis: U.S. Naval Institute, 1969), 66.

8. Francis Smith Belton served through the War of 1812 as a staff officer, and by the outbreak of the Mexican War he had become lieutenant colonel of the Third Artillery. He was brevetted to colonel for his conduct at Contreras and Churubusco.

9. Justin Smith states that "a company of sappers and miners and an iron boat loaded with entrenching tools and sand-bags accompanied Worth's brigade." Justin H. Smith, *The War with Mexico,* 2 vols. (New York: Macmillan, 1919), 2:336n16.

10. The wooden steamer *Massachusetts* was constructed in Boston in 1845 and helped pioneer commercial steamer service between New York and Liverpool. It was purchased by the War Department in 1847 and served as a troop transport.

11. James P. Dickenson was the lieutenant colonel of Col. Pierce M. Butler's Palmetto Regiment, South Carolina Infantry. He was mortally wounded at the Battle of Churubusco.

12. The escopeta was a .69-caliber musket popularized in the mid-seventeenth century by Spanish cavalry on the colonial frontier. According to one observer, the weapon was a "short bell-mouth, bull-doggish looking musket, carrying a very heavy ball, which is 'death by law' when it hits, but that is seldom, for they shoot with little accuracy. They are good for nothing except to make a noise." John C. Duval, *The Adventures of Big-Foot Wallace* ([Macon, Ga.]: J. W. Burke, 1870), 173.

13. Francis Taylor, class of 1825, was brevetted to major in the Fourth Artillery for his conduct at the Battle of Cerro Gordo and to lieutenant colonel for his service at Churubusco.

14. Pierre Gustave Toutant Beauregard, the son of an old Louisiana creole family, was graduated from West Point, ranking second in the class of 1838, and was commissioned into the Second Artillery but soon transferred to the Corps of Engineers. Among Scott's engineers in Mexico, he highly distinguished himself, earning brevets to captain for Contreras and Churubusco and to major for Chapultepec. Stevens considered Beauregard to be "one of the finest soldiers in our corps. Of great strength, accomplished in all manly exercises, well read in his profession, and of forcible and independent character, much self-reliance and confidence, he has established a good reputation throughout the service." In January 1861 Beauregard was appointed as superintendent of the U.S. Military Academy but resigned after Louisiana's secession, later that same month, to become one of the South's

leading and most controversial commanders. Beauregard, *With Beauregard in Mexico;* Stevens, *Life of Isaac Stevens,* 216–17; T. Harry Williams, *Beauregard: Napoleon in Gray* (Baton Rouge: Louisiana State University Press, 1955).

15. Joseph Gilbert Totten was graduated third in the West Point class of 1805 and was commissioned into the engineers. During the War of 1812, he was chief engineer on the staff of Maj. Gen. Stephen Van Rensselaer and was brevetted to lieutenant colonel for his conduct during the Battle of Plattsburg. He was appointed chief engineer in 1838 with the rank of colonel, serving in that position for twenty-five years. For his service at the siege of Vera Cruz, Totten was brevetted to brigadier general.

16. John Gray Foster was graduated fourth in McClellan's West Point class of 1846 and was commissioned into the topographical engineers. During the Mexican War he earned brevets for "gallant and meretricious conduct" first at Contreras and Churubusco and then at Molino del Rey.

17. McClellan's pique regarding this point is puzzling in that Totten's report to Scott lauded the engineers for their "merits and services" during the siege and cited the sappers and miners in particular, mentioning Swift, Smith, and McClellan by name. Later Totten wrote directly to William L. Marcy that the officers of the company "participated in these night and day labors" and directed the operations of the siege with such "unsurpassed intelligence and zeal" that they were "emulated in perseverance and devotion by these good and faithful soldiers." Joseph G. Totten to Winfield Scott, 28 Mar. 1847, in *Report of the Secretary of War,* 30th Cong., 1st sess., 1847, H. Exec. Doc. 8, 244–45; Totten to Marcy, "Report of the Chief Engineer," in *Report of the Secretary of War,* 31st Cong., 1st sess., 1849, H. Exec. Doc. 1, 279–80.

18. John Lind Smith, the senior engineer on Scott's staff after the departure of Joseph Totten and until relieved by Robert E. Lee after the capture of Mexico City, was commissioned directly into the engineers during the War of 1812. He received brevets through the rank of colonel "for gallant and meritorious conduct" during the Mexico City Campaign.

19. Although the landing met with no resistance, Scott remained concerned that the onset of the yellow-fever season would catch his army in the low country and destroy it more surely than could Mexican guns. At the same time, he wished to avoid the high casualty rate that storming the walls of Vera Cruz would entail. Accordingly on 22 March he began the systematic bombardment of the city. The Mexican commandant, Juan Morales, resigned in favor of Gen. of Brigade José Juan de Landero, who on 27 March agreed to surrender the city and castle and their combined garrisons, numbering 4,000 men.

Diary Entry

VERA CRUZ

14 MARCH 1847

The next day Foster was detailed to assist Major Smith and Beauregard in measuring a base line etc. on the sand hills. G. W. and myself went to the lime kiln in the morning, where we saw Captain Vinton,[1] Van Vliet,[2]

Loring,[3] Rodgers[4] and Wilcox[5] (Cadmus)—took a good look at the town and its defences—and determined to go along the ridge by the cemetery that night and to go nearer the city.

While at the lime kiln an order was received from General Worth informing Captain Vinton that the enemy's picquets would be driven in that day and that he (Captain Vinton) must not attempt to support them—as there were strong reserves.

We returned to camp, got our dinner and started again—being a little fearful that our picquets would be so far advanced as to interfere with our operations. But we found them about 150 yards in advance of the line of investment, stooping, whispering, and acting as if they expected to be fired upon every moment—whilst we had been a mile and a half in advance of their position with a dozen men. They were at first disposed to dissuade us from going on—as being too dangerous, etc. We went on though, accompanied by Captain Walker of the 6th.[6] The Captain left us before we got to the cemetery.

I took one man (Sergeant Starr)[7] and went down to reconnoitre it—in order to ascertain whether it was occupied by the enemy, whilst G. W. went on to examine a hill which covered the valley from Santiago and the Castle to some extent. I went down to the cemetery (finding a good road) went around it and got in it—satisfying myself that it was not occupied. I rejoined G. W. and together we went on very near the town. We returned late, being the only officers of any corps who had gone as far as, much less beyond the cemetery.

1. John Rogers Vinton, class of 1817, was brevetted to major in the Fourth Artillery for his service at the Battle of Monterrey but was killed on 22 March 1847 "by the wind of a shell" during the siege of Vera Cruz. Winfield Scott wrote of Vinton that he was "one of the most talented, accomplished, and effective members of the army" and "was highly distinguished in the brilliant operations at Monterey." Scott to Marcy, 23 Mar. 1847, in Edward D. Mansfield, *The Mexican War* (New York: A. S. Barnes, 1849), 178; John Adams Vinton, *The Vinton Memorial* (Boston: S. K. Whipple, 1858), 284.

2. Stewart Van Vliet, class of 1840, spent the largest part of his army career in the Quartermaster Department, mostly on assignments on the western frontier. During the Civil War he served under McClellan as chief quartermaster of the Army of the Potomac.

3. In 1846 William Wing Loring was commissioned a captain in the newly formed Regiment of Mounted Riflemen and was engaged in Scott's campaign from Vera Cruz to Mexico City. On 16 February 1847 he was placed in command of the Mounted Riflemen and

promoted to the rank of major. Loring was brevetted to lieutenant colonel for his role in the Battle of Contreras and to colonel for Chapultepec and the Belen Gate, where he received a wound that required the amputation of his left arm.

4. Alexander Perry Rodgers, class of 1846, was a second lieutenant in the Fourth Infantry when he was killed in action at the storming of Chapultepec.

5. Cadmus Marcellus Wilcox, class of 1846, served as adjutant of the Fourth Infantry and as aide de camp to John A. Quitman. He was brevetted to first lieutenant for his conduct in the Battle of Chapultepec. In 1860 he served as senior groomsman at the wedding of his friend McClellan to Ellen Mary Marcy. Wilcox was the author of the posthumously published *History of the Mexican War* (Washington, D.C.: Church News Publishing, 1892). Gerard A. Patterson, *From Blue to Gray: The Life of Confederate General Cadmus M. Wilcox* (Mechanicsburg, Penn.: Stackpole, 2001), 12.

6. William Henry Talbot Walker, class of 1837, was brevetted to major for his conduct during the battles of Contreras and Churubusco and to lieutenant colonel for his role in the Battle of Molino del Rey, where he received a severe wound.

7. Samuel Henry Starr was born in Ireland in 1813. In 1832, after immigrating to the United States, he enlisted in Company G, Fourth U.S. Artillery, rising to the rank of sergeant in 1837. For his exemplary service in the engineering company, he was brevetted a second lieutenant in the Second Dragoons on 28 June 1848. With the outbreak of the Civil War, Starr was appointed colonel of the Fifth New Jersey Infantry but resigned from the volunteers to become major of the Sixth U.S. Cavalry, 25 April 1863. Less than three months later, he was severely wounded and captured when his regiment was virtually destroyed in a cavalry action at Fairfield, Pennsylvania, during the Gettysburg Campaign. Nevertheless, following his exchange, Starr was promoted to colonel, at which rank he retired on 15 December 1870. Guy V. Henry, *Military Record of Civilian Appointments in the United States Army* (New York: D. Van Nostrand, 1873), 182; William Harding Carter, *From Yorktown to Santiago with the Sixth Cavalry* (Baltimore: Lord Baltimore, 1900), 91–97.

Diary Entry

VERA CRUZ

15 MARCH 1847

The next day we were ordered to cut an infantry road as far as the cemetery. We found that one had been cut before we got out by Captain Johnston[1] as far as the old grave yard. We cut one completely concealed from view from there to the hollow immediately opposite the cemetery.

Captain Walker's company was behind the cemetery. Whilst there one of his sentinels reported the approach of some Lancers. They stopped at a house about thirty yards from the other side of the cemetery—and came no farther. On the strength of the approach of these fifteen or twenty Lancers a report got back to camp that the advanced picquets had been attacked by a strong force of Mexicans—so on our return we met nearly the whole division marching out to drive them back—litters for the "to be wounded" and all. It was a glorious stampede—well worthy of Bold Billy Jenkins [i.e., William Jenkins Worth].

1. Joseph Eggleston Johnston, class of 1829, received two serious wounds during the Battle of Cerro Gordo but upon recovery was transferred to the Voltigeur Regiment with a promotion to lieutenant colonel. He was brevetted to colonel for "gallant and meritorious conduct" at Cerro Gordo and Chapultepec.

When Johnston was again wounded at Chapultepec, General Scott observed that "Johnston is a great soldier, but he has an unfortunate knack of getting himself shot," and his West Point classmate, Robert E. Lee, commented that "Joe Johnston is fat, ruddy, and hearty. I think a little lead, properly taken, is good for a man." Fifteen years later during the Peninsula Campaign, he led Confederate forces against McClellan's Army of the Potomac until he was severely wounded during the Battle of Seven Pines, 31 May 1862. Johnston was replaced in command by Gustavus Woodson Smith, who was in turn superseded by Robert E. Lee. Robert E. Lee to John Mackay, 2 Oct. 1847, Robert E. Lee Papers, U.S. Army Heritage and History Center, Carlisle Barracks, Pa.; Winfield Scott, *Memoirs of Lieut.-General Winfield Scott, LL.D.* (New York: Sheldon, 1864), 184; Craig L. Symonds, *Joseph E. Johnston: A Civil War Biography* (New York: W. W. Norton, 1992).

Diary Entry

VERA CRUZ

16–18 MARCH 1847

The next day we went out [and] met Major Scott[1] who went with G. W. to [the] position afterward occupied by the six gun battery—whilst I had a hole made through the cemetery wall and broke into the chapel—hoping to be able to reach the dome, and ascertain from that place the direction of the streets. I could not—we rather—get up to the dome, so we left the cemetery, determining to push on toward the town. G. W. found a very fine position for a battery about 450 yards from Santiago and enfilading the principal street. We met Colonel Totten and Captain Lee—showed them the place—they were very much pleased with it.

We came out with the Company (Captain Lee, Smith, Foster and myself) that evening, arrived at the place after dark, and Captain Lee, Smith and Foster went in to lay out a battery—leaving me, in command of the Company, in the road. When on our return we were passing by the old grave yard a sharp fire of musketry commenced—one of our pickets had been fired upon.

The next day (17th) we cut a path to the position of this battery (in perspective). As we returned they discovered us and opened a fire of 24-pound shot upon us which enfiladed our path beautifully. They fired too high and hit no one. We reached at length a sheltered position where we remained until the firing ceased—the balls striking one side of the hill—we being snugly ensconced on the other.

On the next day (18th) the position of the batteries was definitely fixed. In the afternoon I was ordered by Colonel Totten to arrange at the Engineers' Depot (on the beach) tools for a working party of 200 men—and be ready to conduct it as soon as it was dark to the proper position. The working party (3rd Artillery, Marines, and 5th Infantry—all under Colonel Belton) did not arrive until long after dark—and it was quite late when we arrived at the position for the batteries.

I was placed in charge of Mortar Battery No. 1—G. W. in charge of No. 2—a parallel was also made across the little valley. Each of these batteries was for three mortars. No. 1 was formed by cutting away the side of a hill, so that we had merely to form the epaulments and bring the terreplein down to the proper level—the hill sheltering us from the direct fire of the Castle and Santiago. So also with No. 2—which was made in the gorge where the road to the cemetery crossed the ridge on [the] left of [the] valley.

The tools for [the] working party were arranged on the beach in parallel rows of tools for 20 men each and about four feet apart, so that they might take up the least possible space. Each man was provided with a shovel and either a pick, axe, or hatchets (about 140 picks and mattocks). The party was conducted in one rank, by the right flank. The men were well covered by daylight.

1. John Benjamin Scott, class of 1821, was brevetted to major in the Fourth Artillery for his conduct at Palo Alto and Resaca de la Palma.

Diary Entry

VERA CRUZ

19 MARCH 1847

Mason, Foster, and I think Stevens,[1] relieved Captain Lee, Beauregard, Smith and myself at three A.M. During the day they continued the excavation of the two batteries and the short parallel across the valley. The enemy kept up a hot fire during the forenoon but injured no one. During the evening of this day Smith laid out and commenced the parallel leading from No. 1 to the position afterward occupied by the 24-pounder battery. The work was difficult on account of the denseness of the chaparral and the small number of workmen. The parapet was made shot proof (or sufficiently so to answer the purpose of covering the morning relief) by daybreak. The enemy fired grape etc. for a short time, but not sufficiently well timed or long enough kept up to impede the progress of the work.

The battery known as the Naval Battery[2] was commenced on this same night. The enemy were kept in entire ignorance of the construction of this battery until the very night before it opened, and then they only discovered that something was being done there—they did not know what. The Mexican Chief Engineer told Colonel Totten of this fact after the capitulation.

1. James Louis Mason, class of 1836, was brevetted through lieutenant colonel of engineers for his conduct during the Mexico City Campaign. Stevens was of the opinion that no officer of engineers rendered "more brilliant service than Captain Mason. Of remarkable intellectual force, great quickness of apprehension, highly cultivated, an ambitious student, and frank and honest in his life, on the field of battle, in a reconnaissance of the enemy's position, indeed in every emergency, he has been conspicuous for force, rapid decision, and the most daring intrepidity." Stevens, *Life Isaac Stevens*, 217–18.

Isaac Ingalls Stevens was graduated first in the class of 1839. He was commissioned into the Corps of Engineers and was engaged in the construction of coastal defenses in New England. After the siege of Vera Cruz, he was made adjutant to Maj. John L. Smith, the chief engineer of Scott's army. Stevens saw action in all of the battles around Mexico City, receiving brevets through the rank of major as well as a serious wound at Chapultepec. In 1851 he subsidized publication of one thousand copies of a short book, *Campaigns of the Rio Grande and Mexico* (New York: D. Appleton, 1851), a defense of General Scott's role in the war.

2. Commodore Matthew C. Perry, commander of the Gulf Squadron, loaned the army three long 32-pounders and three 8-inch Paixhain guns. Under the supervision of Robert E. Lee, the Naval Battery was erected some seven hundred yards south of the Vera Cruz wall

on the night of 23 March. "Our friends of the navy," wrote Scott, were "unremitting in their zealous co-operation, in every mode and form." The naval cannon, both ponderous and precise, produced a fifty foot-wide breech in the city's wall, terrorizing and demoralizing the garrison.

Diary Entry

CAMP NEAR VERA CRUZ

20 MARCH 1847

The construction of the parallel and of the mortar batteries Nos. 1 and 2 was carried on during this day. By three P.M., when Mason and myself went out there the parallel was finished, the excavation of the two batteries completed, the sandbag traverses in No. 2 finished, those in No. 1 very nearly so. We were to lay out and excavate the positions for the two magazines of each battery, to commence Mortar Battery No. 3 (for four mortars), lay the platforms and place the magazine frames—which were to be brought out at night fall.

By the direction of Mason, I had the positions of the magazines prepared and laid out before dark. Colonel Totten came out and directed me to lay out No. 3. I also laid out the boyau leading from 1 to 2. Mason took charge of the magazines of 1 and 2 and directed me to take charge of No. 3. I employed four sets of men on the battery at the same time—one set throwing the earth from the rear of the parallel upon the berm, a second on the berm disposing of this earth thrown on the berm, a third set working at the rear of the battery, excavating toward the front (these threw the earth so as to form slight epaulments), and in rear. A fourth set were employed in making the excavations for the magazines. A very violent Norther arose which obliged me to employ the first and second sets in front of the battery—they excavating a ditch.

At daylight the parapet was shot proof and the battery required about one hour's digging to finish it. Owing to some mistake, the platforms and magazine frames did not arrive until very late and but little progress was made as far as they were concerned. Had they arrived in time all three batteries could have opened on the afternoon of the 21st. The construction of the battery on the left of the railroad still progressing. They fired rockets etc. at us during the early part of the night.

Diary Entry

CAMP NEAR VERA CRUZ

21 MARCH 1847

During this day not very much was done—some progress was made with the six gun battery—magazines, platforms, etc.

Diary Entry

CAMP NEAR VERA CRUZ

22 MARCH 1847

Not being aware of a change in the detail, I went out at three A.M. Found the magazines of No. 2 finished, the small magazines of No. 1 the same. Took charge of large magazine of No. 1—whilst Mason was engaged with those of No. 3. About eight [o'clock] was informed of change of detail, went to camp and was requested by Colonel Totten to go out to the trenches "extra" and give all the assistance in my power, since the General wished to send in a summons to the town at two P.M. and open upon them if they refused to surrender.

I went out and was chiefly occupied during the day in covering the magazine of No. 1 with earth. This was done under fire of Santiago and adjacent bastion, which batteries having a clear view of my working party made some pretty shots at us—striking the earth on the magazine once in a while, but injuring no one. At two P.M. we were ready to open with three mortars in No. 1—three in No. 2—one in No. 3—seven in all.

The flag was carried in by Captain [Joseph E.] Johnston, the enemy ceased firing when they saw it. Colonel Bankhead[1] informed the Commandants of Batteries 1 and 3 that the discharge of a mortar from No. 2 would be the signal to open from all the mortars. The flag had hardly commenced its return from the town when a few spiteful shots from Santiago at my party on the magazine told us plainly enough what the reply had been. Probably half an hour elapsed before a report from No. 2 gave us the first official intimation that General Morales[2] had bid defiance to us, and invited us to do our worst.

The command "Fire!" had scarcely been given when a perfect storm of iron burst upon us. Every gun and mortar in Vera Cruz and San Juan, that could be brought to bear, hurled its contents around us—the air swarmed with them—and it seemed a miracle that not one of the hundreds they fired fell into the crowded mass that filled the trenches. The recruits looked rather blue in the gills when the splinters of shells fell around them; but the veterans cracked their jokes and talked about Palo Alto and Monterrey.

When it was nearly dark I went to the left with Mason and passed on toward the town where we could observe our shells—the effect was superb. The enemy's fire began to slacken toward night, until at last it ceased altogether—ours, though, kept steadily on, never ceasing—never tiring.

Immediately after dark I took a working party and repaired all the damage done to the parapets by the enemy's fire, besides increasing the thickness of the earth on the magazines of No. 1. Captain Vinton was killed a short time before dark near Battery No. 3 by a spent shell. Two men were wounded by fragments of shells near No. 1.

Shortly after dark, three more mortars were put in Battery No. 3—making 10 mortars in all. Captain [John L.] Saunders was employed upon the 6 gun battery (24-pounders). He revetted it with one thickness of sand bags, all of which fell down next morning. I brought out from the Engineer Depot the platforms for this battery during the night—the magazine frame was brought out next day. The battery on the left of the railroad still progressing, under the charge of Captain Lee, Tower,[3] and [G. W.] Smith—who relieved each other.

1. James Bankhead was commander of the Second Artillery and Winfield Scott's chief of artillery during his campaign against Mexico City, earning a brevet to brigadier general for his conduct during the siege of Vera Cruz.

2. Juan Morales, as colonel of the Permanent Reserve Infantry Battalion of San Luis, played a principle role at the siege and storming of the Alamo during the War for Texas Independence. Promoted to general of division, Morales was appointed commandant general of Vera Cruz in 1847. Rather than surrender the city and its garrison to Winfield Scott, he turned the command over to José Juan de Landero and fled the doomed city by boat on the night of 25–26 March 1847. Santa Anna ordered Morales to be tried by court-martial, but he died in 1847 before action against him could be taken.

3. Zealous Bates Tower graduated at the head of the class of 1841 and was commissioned in the Corps of Engineers. In 1842 he returned to West Point as an assistant professor of engineering. During the Mexico City Campaign, he served on Scott's staff, earning brevets through major. Isaac Stevens opined that Tower, "for judgment, for an assured and natural self-reliance, great force of character, and great decision and intrepidity in emergencies, has no superior in our corps." Stevens, *Life of Isaac Stevens*, 217.

Diary Entry

CAMP NEAR VERA CRUZ

23 MARCH 1847

Firing continued from our mortars steadily—fire of enemy by no means so warm as when we opened on the day before. Our mortar platforms were much injured by the firing already. The 24-pounder battery had to be re-revetted entirely—terreplein leveled. During this day and night the magazine was excavated and the frame put up. Two traverses made—the positions of platforms and embrasures determined. Two platforms laid and the guns run in—the embrasures for them being partly cut. One other gun was run to the rear of the battery.

Diary Entry

CAMP NEAR VERA CRUZ

24 MARCH 1847

On duty with Captain Saunders again—could get no directions so I had the two partly cut embrasures marked with sand bags and dirt, and set a party at work to cover the magazine with earth as soon as it was finished. During this day the traverses were finished, the platforms laid, the magazine entirely finished, and a large number of sand bags filled for the revetments of the embrasures.

The "Naval Battery" opened today, their fire was fine music for us, but they did not keep it up very long. The crash of the eight inch shells as they broke their way through the houses and burst in them was very pretty. The "Greasers" had had it all in their own way—but we were gradually opening on them now.

Remained out all night to take charge of two embrasures. The Alabama Volunteers,[1] who formed the working party, did not come until it was rather late—we set them at work to cut down and level the top of the parapet—thickening it opposite the third and fourth guns. Then laid out the embrasures and put seven men in each. Foster had charge of two, Coppée[2] of two, and I of two. Mine were the only ones finished at daylight—the Volunteers gave out and could hardly be induced to work at all.

1. On this day Pvt. Stephen F. Nunnalee of Company D, First Alabama Volunteers wrote in his diary that "twenty men out of the company [were] detailed to work on our batteries." Some were detailed to "work in the trenches, fill sand bags, and construct forts," Nunnalee complained, while others were put to work hauling into camp provisions deposited on the beach by the navy. Stephen F. Nunnalee, "Letter to Dr. W. S. Wyman from S. F. Nunnalee," *Alabama Historical Quarterly* 19 (1957): 425; Butler, *Eutaw Rangers*, 107.

2. Henry Coppée, class of 1846, was brevetted to captain in the First Artillery "for gallant and meritorious conduct" in the battles of Contreras and Churubusco.

Diary Entry

CAMP NEAR VERA CRUZ

25 MARCH 1847

Mason and Stevens relieved Beauregard and Foster—but I remained. I had the raw hides put on and with a large party of Volunteers opened the other embrasures. This was done in broad daylight, in full view of the town—yet they had not fired more than three or four shots when I finished and took in the men. The battery then opened.

We then gave it to [the] Mexicans about as hotly as they wished. We had ten mortars—three 68s, three 32s, four 24s, and two eight-inch howitzers playing upon them as fast as they could load and fire. Captain Anderson,[1] 3rd Artillery, fired on this morning thirty shells in thirty minutes from his battery of three mortars (No. 1).

As I went to our camp I stopped at Colonel Totten's tent to inform him of the state of affairs—he directed me to step in and report to General Scott. I found him writing a despatch. He seemed to be very much delighted and showed me the last words he had written which were "indefatigable Engineers."

Then we were needed and remembered—the instant the pressing necessity passed away we were forgotten. The echo of the last hostile gun at Vera Cruz had not died away before it was forgotten by the Commander in Chief that such a thing existed as an Engineer Company.

The superiority of our fire was now very apparent. I went out again at three P.M.—met Mason carrying a large goblet he had found in a deserted ranch. Found Captain Lee engaged in the construction of a new mortar battery for four mortars, immediately to the left of No. 1—in the parallel. There was a complete cessation of firing—a flag having passed in relation to the consuls, I think.

The platforms of this battery were laid, but not spiked down. A traverse was made in boyau between Nos. 1 and 2, just in front of the entrance of the large magazine of No. 1, it being intended to run a boyau from behind this traverse to the left of the new battery. I laid out a boyau connecting Stevens's communications with the short "parallel" of No. 2, then Captain Lee explained his wishes in relation to the new battery and left me in charge of it. I thickened the parapet from a ditch in front—inclined the superior slope upward, left the berm, made the traverses, had the platforms spiked, etc. The mortars were brought up and placed in the battery that night. Captain Saunders sent me to repair the embrasures of the 24-pounder battery—doing nothing himself. He then sent me to excavate the boyau I had laid out.

About 11:30 the discharge of a few rockets by our Rocketeers caused a stampede amongst the Mexicans—they fired escopettes and muskets from all parts of their walls. Our mortars reopened about 1:30 with the greatest vigor—sometimes there were six shells in the air at the same time. A violent Norther commenced about one o'clock making the trenches very disagreeable.

About three quarters of an hour or an hour after we reopened we heard a bugle sound in town. At first we thought it a bravado—then reveillé, then a parley—so we stopped firing to await the result. Nothing more was heard, so in about half an hour we reopened with great warmth. At length another *chi-wang-a-wang* was heard which turned out to be a parley. During the day the terms of surrender of the town of Vera Cruz and castle of San Juan de Ulua were agreed upon, and on 29th of March, 1847 the garrison marched out with drums beating, colors flying and laid down their arms on the plain between the lagoon and the city [] muskets were stacked and a number of escopettes [] pieces of artillery were found in the town and [] in the castle.[2]

1. Robert Anderson, best known as the defender of Fort Sumter, was graduated from West Point in 1825 and commissioned into the Third Artillery. He received a major's brevet for his role in the Battle of Molino del Rey. His Mexican War correspondence was published as *An Artillery Officer in the Mexican War* (New York: G. P. Putnam's Sons, 1911).

2. McClellan not infrequently left blank spaces in his diary and returned to them later to fill in missing information. In this particular case he did not supply the missing numbers.

George B. McClellan to Arthur and Mary McClellan

ENGINEER CAMP NEAR VERA CRUZ

2 APRIL 1847

Benditos niños mios,

My dear little children,

If the "burros" here do not deafen me by their continual braying, so that [I] can't *see* to write, I will tell you what kind of place I am in. Our Campo is on the beach, the sea is not thirty steps from me—so that I have plenty of fine sea bathing. There are two objections to going out too far, one is that you will get on the coral reef and cut your feet, and the other that if you cross the reef you will make rather too intimate an acquaintance with certain shovel-nosed gentlemen that abound in these regions, and which you people "way up North, about New Orleans, Philadelphia, &c." call sharks.

When sailing down to this place from Tampico, I used to amuse myself by trying to harpoon the rascals. There are plenty of dolphins in this gulf—they are a beautiful blue color when in the water, and when caught and dying on the deck of a vessel change their colors most beautifully. Have you ever read about the salamander, which lives on air and changes its color so often? I have seen them in Mexico. They have a comb on their heads, like those which ladies wear in their heads, and change color very often.

When our schooner came to Anton Lizardo we got there about dark, we were sailing along very smoothly and nicely when "slam," we were on a coral reef—we were in a nice predicament you may well imagine. The skipper put out an anchor astern by means of his little boat and the men pulled in the cable so as to bring the schooner to where the anchor was, but the cable broke and the anchor was lost. Just about that time some boats belonging to the frigate *Raritan* came along. One of them took me on board where I saw the Commodore (Conner) and told him how [we] were placed. He ordered off some boats to put us on a little desert Island which was there, and ordered a little steamer, called the *Spitfire,* to pull off the schooner.

The boat which took me to the Island was so heavy that it could not get nearer than 50 yards to the shore, so one of the sailors took me on his shoulders and carried me on shore. Some of the sappers thought they would try the same thing, but just as I reached the shore I heard some splashes, then some loud laughing. I turned round and saw that the "Jack Tars" had "spilled" my sappers, or as they said, "hove 'em overboard." I had a good laugh at the trick.

I went in the water next morning, and when I got in Coral Reef it was a beautiful sight. There were pretty pieces of red and white coral, all kinds of sea plants growing there, and beautiful fish swimming around. The water was clear as crystal.

The beach is covered with wagons, boats, tents, jackasses, Irishmen, Dutchmen, and every thing you can think of.

Write me a letter and tell me whether you want a parrot. Tell me what you wish me to bring you when I come home, and if it is possible, I will bring you whatever you want.

You must have plenty of ice for me when I come back. I suppose you are in the middle of snow and ice now, but it is so hot here that I can hardly live—and it will soon be still hotter. I suppose I will find you, Arthur, a young gentleman with a frock coat &c., and you, Miss Mary, a grown up young lady when I come back—too proud to know your brother who has been "roughing it" so long in the wars.

Good bye my dear children and write (if you can) to your brother,
Geo. B. McC

5

"NOTHING SEEMED TO THEM TOO BOLD"

CERRO GORDO TO MEXICO CITY

26 March 1847–30 May 1847

During the days following the surrender of Vera Cruz, McClellan was engaged in making surveys of the captured fortifications, dismantling the army's siege works, and landing and organizing the engineers' train. He and his company had to work quickly, however, for the yellow-fever season was almost upon them, and so on 12 April 1847 the company of sappers and miners departed the coast with General Scott and his staff.[1]

Five days later McClellan and the engineer company reached Cerro Gordo, the strategically vital pass that was the gateway to the Mexican highlands and ultimately to Mexico City. There President and Generalissimo Antonio López de Santa Anna had fortified the naturally strong positions on El Telegapho and La Atalya in anticipation of halting the Yankee invaders on the fever-ridden coastal plain.

In accordance with intelligence gathered during a brilliant reconnaissance mission by Robert E. Lee, Scott determined to hold Santa Anna's forces in place with a feint along his front by General Pillow's brigade of volunteers while turning the Mexican left flank and cutting its line of retreat with his regulars under General Worth.

Much to his disgust, McClellan was attached to the volunteers, to whom he derisively referred as the Duck Creek Fencibles. Placed in charged of a ten-man detachment from the engineer company, he was directed to clear the obstacles between Pillow's command and the Mexican line so that the volunteers could carry out their attack. "This was a service of no common danger," wrote his first biographer, George Stillman Hillard, "as the heavy and well-served Mexican batteries in front swept the space before them with a most destructive fire."[2]

Although Pillow totally bungled his orders and needlessly sacrificed the lives of many men of his brigade, the American victory at Cerro Gordo was brilliant and absolute—so much so that fourteen years later McClellan, as commander of the Department of Ohio, attempted to employ Scott's tactics at Rich Mountain, Virginia. Recalling the useless sacrifice of Pillow's men in a headlong assault on enemy entrenchments, McClellan vowed to "repeat the manoeuvre of Cerro Gordo" and assured Scott, who was still the general in chief of the U.S. Army, "that no prospect of a brilliant victory shall induce me to depart from my intention of gaining success by manoeuvring rather than by fighting; I will not throw these men of mine into the teeth of artillery and intrenchments, if it is possible to avoid it."[3]

As at Vera Cruz, the engineers were singled out for high praise for their signal contribution to the victory at Cerro Gordo. Despite the icy contempt of the West Point–trained regulars, Pillow praised McClellan, reporting that he and Zealous Bates Tower had "displayed great zeal and activity in the discharge of their duties in connection with my command."[4]

Scott, in his report to Marcy, singled out McClellan, Smith, Beauregard, Stevens, and Tower as having "won the admiration of all about them." Stevens, writing to his wife, declared that "General Scott expressed himself in terms that won my heart. He remarked, 'You engineers are too daring. You require to be held back.'"[5]

But the debacle on Pillow's front further solidified McClellan's already ingrained prejudice against volunteers, especially volunteer officers and officers appointed to the regular army directly from civilian life. He considered the commanders of the new regiments "deficients of the Military Academy, friends of politicians, & bar room blackguards." This too was an attitude that McClellan would maintain through the Civil War.

Having dispersed Santa Anna's army and secured the pass at Cerro Gordo, Scott pushed on toward Mexico City, with the leading division under Twiggs reaching Jalapa on 20 April. There Stevens, as he wrote in his diary, met Smith and McClellan, who had arrived with the engineer company in the vanguard of the army. Much of their equipment had been left behind, however, due to lack of transportation, "the mules having proved very poor on the route from Plano del Rio, and many of the animals being entirely unserviceable." Lieutenant Foster was sent back to Plan del Rio to bring up eight wagonloads of engineering implements, and in the meantime McClellan remained "busily engaged in organizing the train of the engineer company." While thus occupied, however, McClellan had the consolation of receiving from the War Department on 24 April his promotion to the substantive grade of second lieutenant.[6]

On 26 April the sappers and miners moved out of Jalapa as the vanguard of Worth's division, marching toward the Mexican capital.[7] As the army neared Puebla, rumors of enemy resistance became increasingly insistent. On 13 May at the village of Nopalucan, Scott received word that Santa Anna, with some fifteen hundred or two thousand lancers, had passed through there three days earlier bound for Puebla and that they had "prepared mines in the road at El Pinal." In consequence the engineer company, supported by Capt. James Duncan's battery of the Second Artillery and a detachment of infantry, was dispatched to examine the pass. But the mines, "of which rumor was so big, were little excavations commenced under the road in two places," reported Stevens.[8]

Later that same day McClellan, with the vanguard under Worth, arrived at Amosoque, a small town twelve miles west of Puebla. "Our officers did not dream of finding any portion of the enemy here," he later told Hillard, "and the usual precautions adopted to guard against surprise were somewhat relaxed."[9]

On the morning of 14 May, the soldiers were busily cleaning their arms and equipment "in order," wrote Hillard, "that they might enter Puebla in good train." At about nine o'clock, however, a drummer boy who had wandered out of the village raced back, sounding the alarm that a column of lancers was rapidly coming their way. A hasty reconnaissance led by Lee estimated their number at fewer than two thousand, but the long roll called Worth's regiments into line of battle, and McClellan, who was quartered on the side of town from which the enemy was approaching, quickly mounted and rode out to observe the oncoming lancers.

After riding a few hundred yards, he later told Hillard, McClellan encountered a captain of Mexican cavalry conducting a reconnaissance of the town. Neither officer was escorted, and both were armed with pistols and sabers. The Mexican turned and fled, but McClellan, mounted on the blooded mare that had been a gift from his Alabama brother-in-law, overtook him and forced him to surrender. On their return to the village, however, the Mexican officer suddenly "wheeled his horse about and raced away" while attempting to draw his pistol. McClellan, according to Hillard, recaptured the lancer and "gave him to understand that if he renewed the attempt to escape, he should be obliged to put a bullet through him." They then rode "triumphantly" back to headquarters, where McClellan "surrendered his prisoner to his commanding officer."[10]

Stevens's account of this adventure, however, is at variance with that which McClellan told to Hilliard. According to a letter that Stevens wrote to his wife shortly after the incident—and much earlier than McClellan's 1864 campaign biography—the Mexican cavalry was checked by the guns of Duncan's and Steptoe's batteries and retired without offering a serious threat. Somewhat rashly perhaps, McClellan, Stevens, and three dragoons followed the retreating lancers to the hacienda San Miguel, some five miles from Amasoque, but losing sight of the fleeing column, the five Americans returned to their own lines. In route, however, they encountered a captain of Mexican cavalry together with his servant.[11]

In Stevens's telling, McClellan "nearly captured" the Mexican officer. Although, he wrote his wife, "McClellan was very gallant and prompt in pursuing the Mexican," the lancer eluded his pursuer in the chaparral, and the five Americans returned to camp emptyhanded.[12]

Following the affair at Amasoque, Scott reached Puebla on 28 May 1847, the route having been reconnoitered by his engineer officers escorted by the sappers. There the army was once again delayed, this time for thirteen weeks while awaiting reinforcements from the States to replace the seven volunteer regiments whose terms of enlistment had expired. Making virtue of necessity, McClellan drilled the engineer company in *The School of the Sapper* and *The School of the Miner* and constructed fortifications by day and by night studied Cortés's conquest of Mexico, three hundred years earlier.[13]

There too McClellan received word of the death of his father, Dr. George McClellan, and "for several days," wrote his friend G. W. Smith, "he would see no one and was inconsolable."[14]

While awaiting reinforcements, Scott ordered his engineers "to collect information in reference to the several routes to the capital, the obstacles of nature and of art which would obstruct our advance, the troops and the material which would be at the disposal of the enemy."[15]

At last, having received a brigade of new recruits, Scott's army left Puebla on 7 August 1847, with McClellan and the engineering company once again in the vanguard. Their assignment was to make the road to Mexico City passable for artillery and wagons. The question was, which road should the army take? Three approaches to the city presented themselves.[16] First was the most direct route by way of the National Road, which had brought the army from Vera Cruz. Second was a detour around Lake Texcoco to the north. The third led around Lakes Chalco and Xochimilco to the south. Scott called upon his engineers to find the optimum route.

With the army halted before El Peñon, an eminence to the east of the city that Santa Anna had fortified to the point of seeming impregnability, Scott dispatched the engineers "to make an ostentatious reconnaissance before the Peñon, with the special charge to run no risk, as the object was to amuse the enemy." The commanding general in fact had no intention of storming the formidable Mexican defenses, but on 13 August he sent several of his engineers, escorted by one company of dragoons and one from the Mounted Riflemen, toward Mexicalzingo, some seven miles to the rear of the Mexican position.[17]

On their ride out McClellan, Beauregard, Capt. Edward Richard Sprigg Canby (Bvt. Brig. Gen. Bennet Riley's assistant adjutant general), and two or three of Sibley's dragoons were attacked by an estimated twenty-five lancers, "well equipped and caparisoned," as Beauregard described them. The five or six Americans "immediately charged upon them at great speed, but before we had reached them, the Mexicans turned about and by hard spurring and yelling, managed to get across a *barranca* or ravine and to reach a small village near by, which saved them from being captured."

Once beyond fear of being made prisoners, however, they turned and opened fire upon the pursuing party. "As the main column of our escort was several miles from us," wrote Beauregard, "we thought it prudent not to pursue them any farther." Interestingly Hillard states that on this occasion McClellan was "saved from probable death or captivity at the hands of about a dozen Mexican lancers by Lieutenant Beauregard and three dragoons," but neither McClellan's letters nor Beauregard's diary mention this incident.[18]

That evening Mason reported to Scott that the route the engineers had explored was practicable and that it could be used to flank Santa Anna out of Peñon. Stevens, the pen of the engineers, declared "this bold movement, almost under the guns of the Peñon, and extremely hazardous in presence of an enterprising enemy was accomplished in the most successful manner." According to Hillard, Scott pronounced the mission "the most daring reconnaissance of the whole war."[19]

Accordingly the army moved out of Ayotla on the morning of 15 August, with the sappers and miners leading the march. In response Santa Anna abandoned his fortifications at El Peñon and shifted his front to face the Americans south of Lake Chalco, occupying a formidable line extending from Mexicalzingo through Churubusco to Contreras. Scott determined to turn the Mexicans out of this position by seizing the village of Contreras on Santa Anna's right.[20]

The engineering company was ordered, "with its tools on the backs of mules," to lead Twiggs's division around the Mexican flank. As the company moved out, McClellan overheard two soldiers of Worth's division conclude that they would not see a fight that day because the engineers were marching away. "I consider it the handsomest compliment that could be paid to the engineer company," McClellan commented to Smith. "The private soldiers of this army understand that we are sent where the hardest work and hardest fighting are to be done—and always at the head of the leading division."[21]

However correct McClellan might have been, the two soldiers were as wrong as men could be, for 19 August 1847 was to witness the first of two days of fighting in two of the most decisive battles in U.S. military history—Contreras and Churubusco.

Upon reaching Twiggs's division at San Augustine later that morning, McClellan took command of the engineer company in G. W. Smith's temporary absence. There, under the direction of Robert E. Lee—who in perhaps the most daring reconnaissance missions of the war, had scouted the route on the day previous—began opening the road to Contreras, "a mule path, requiring to be worked to be practicable for artillery."

The sappers and miners completed a workable route as far as the edge of the Pedregal, a fractured lava flow that rendered passage next to impossible. There the men packed away their tools, "no longer able to work the way in consequence of having come within range of the enemy's batteries," and edged forward, reconnoitering the position of Gen. of Division Gabriel Valencia. G. W. Smith reported hearing sharp musket fire in the direction of the Mexican army and shortly thereafter to have seen Capt. John McClellan and Lt. George B. McClellan riding back toward the American line. The men had come upon the enemy's picket line and drawn heavy fire, and Lieutenant McClellan's horse had been shot from under him.[22]

Valencia's 5,500-man Army of the North was found to be in a strongly entrenched camp on the opposite side of a deep ravine, further protected by the lava bed, and as Stevens later wrote, "the engineers, who advanced close to the enemy's pickets . . . could discover no other route than the mule-path, completely commanded by the long guns of the intrenched camp."

Nevertheless, the divisions of Pillow and Twiggs were ordered to attack the enemy left, and field artillery was called forward to support the assault. "Everything seemed to go wrong," wrote G. W. Smith. Mexican counterbattery fire was intense, sending grapeshot at point-blank range. First Lt. Franklin Dryer Callender's battery, deployed in an exposed position, was "entirely cut up" and its commander shot through both legs. With the engineer company attached to the Third Infantry and serving as combat infantry, McClellan, who had assisted in carrying Callender to the rear, took command of the guns.[23]

For a frightful while his position was perilous, with the battery's supply of spherical case shot running out and its infantry support reduced to some eight rifles as the enemy's skirmishers advanced. Fortunately for McClellan, Persifor Smith's brigade came to his relief at the critical moment.[24]

Night fell on 19 August with the outcome very much in doubt, but on the following morning, General Smith enveloped Valencia's army and destroyed it as an effective fighting force. An infuriated Santa Anna ordered Valencia shot on sight. With the Army of the North, in Stevens's words, "broken to pieces," the troops who had been employed on that front marched rapidly back to the east to deal with the remainder of the Mexican defenders.[25]

The American victory at Contreras had, as Scott had planed, exposed Santa Anna's western flank, forcing the Mexicans to make a determined stand in the fortified village of Churubusco and the nearby San Mateo Convent (persistently referred to in American reports as San Pablo). There McClellan and his company joined Scott's men in time to make "a very rapid reconnaissance" of the field ahead. McClellan and Stevens probed as far as the convent of San Mateo, which the two engineers "observed to be occupied in strength." McClellan, from his advanced position, observed that the roof of the convent was "crowded with troops" and that a battery, largely masked by intervening trees and cornfields, was laid before it.

Once the Mexican position was developed by a further engineer reconnaissance, Smith's brigade, "headed by our gallant engineer company," was launched against the right of Santa Anna's line. A morning of hard fighting, often with bayonets, at last forced the surrender of the convent and the retreat of the remaining Mexican forces from Churubusco, resulting, when combined with the earlier casualties at Contreras, in the loss of some 10,000 men, one third of Santa Anna's total command. Remarkably, although Smith reported that the sappers and miners had been constantly engaged in "clearing away obstacles raised by the enemy, in repairing roads, making bridges, &c., in reconnoitering, and in storming the positions of the enemy" at Churubusco, "no one of my company was touched."[26]

The victorious American army pursued the retreating Mexicans to the suburbs of Mexico City. Rather than follow the routed foe into the capital, however, thus risking the dissolution of the Mexican government, Scott chose to negotiate an armistice with Santa Anna, a truce he hoped would be the precursor of a lasting peace accord between the two North American republics.[27]

Following Contreras and Churubusco, the army was again unstinting in its praise for the company of sappers and miners and its officers. In reporting his role in this action, Beauregard wrote that the behavior of

Smith and McClellan, "as well as that of their sapper-company cannot be too highly praised." G. W. Smith reported that "Lieutenant McClellan, frequently detached, and several times in command of the engineer company, is entitled to the highest praise for his cool and daring gallantry, on all occasions, in the actions of both the 19th and 20th." Maj. John L. Smith, chief engineer, and General Twiggs both mentioned McClellan by name in their official reports, and Isaac Stevens wrote to his wife, "Our gallant engineer company nobly sustained its reputation as the first company in the service." To his diary he further confided, "It seemed to be conceded by the whole army that the engineers in these important operations had done their duty, and that every individual officer had shown a readiness to participate in the perils incident to their service."[28]

But the handsomest praise came from Persifor Smith: "Lieutenant G. W. Smith, in command of the engineer company, and Lieutenant McClellan, his subaltern, distinguished themselves throughout the whole of the three actions" of the nineteenth and twentieth. "Nothing seemed to them too bold to be undertaken, or too difficult to be executed; and their services as engineers were as valuable as those they rendered in battle at the head of their gallant men."[29]

During the armistice that followed these victories, Beauregard wrote, "no reconnaissance was permitted by the Commander-in-Chief, although it was a notorious fact that the enemy was violating it day and night." Thus the company of sappers and miners enjoyed a brief respite from its toils and dangers. Scott's hopes of a permanent peace settlement were dashed, however, and on 6 September he abrogated the armistice and resumed operations against the Mexican capital. The week following proved to be the bloodiest of the war, beginning with the Battle of Molino del Rey and culminating with the capture of Mexico City.[30]

McClellan was not present at Molino del Rey, and though officially tendered a brevet for that action, declined it on the grounds that he was not under fire on that field. His absence there, however, was made up by his performance in the capture of Mexico City.[31]

After seizing Molino del Rey with an excess of blood, the only obstacle standing between Scott's army and the Mexican capital was the old Spanish viceroy's palace, now converted to the national military academy, the castle of Chapultepec. Scott determined to take this key position by storm following an intense artillery bombardment. On 11 September McClellan was placed in charge of a detachment of engineers to construct Battery No. 2, which was to breech the southwestern angle of the castle wall. Following the fall of Chapultepec on 13 September, McClellan was recalled to army headquarters and, with the engineer company, attached to Persifor Smith's division to operate against the San Cosme Garita, one of the main gates to the capital.[32]

On the morning of 13 September, the engineers conducted a reconnaissance of the San Cosme suburb, which revealed that no Mexican artillery opposed an entry in that quarter and that, in Stevens's words, "the infantry force there was not formidable, and the lancers hanging of the flanks were not worthy of regard."[33]

At four o'clock that afternoon, the engineer trains arrived at the garita, and "troops armed with the proper tools, commenced making their way from house to house." Having established itself inside the city wall, the engineering company conducted a reconnaissance toward the center of the capital. Approximately 150 yards beyond the gate, they discovered a fortified convent supported by artillery. G. W. Smith reported this position to General Worth, who sent forward two brigades to clear it. One party, led by the engineer company, reached the top of a three-story building some 40 yards from the battery and opened a plunging fire on the Mexican gunners just as a detachment of the Second Artillery opened a crossfire from the opposite side of the street. As Worth described the scene, the engineer company sprang "as if by magic, to the tops of the houses into which they had patiently and quietly made their way with bar and pick, and to the utter surprise and consternation of the enemy, opening upon him, within easy range, a destructive fire of musketry. A single discharge, in which many of his gunners were killed at their pieces, was sufficient to drive him in confusion from the breastworks." The Mexicans abandoned the battery, carrying away only one of their guns.[34]

Near the end of this engagement, Orderly Sgt. David H. Hastings of the engineer company was shot by a Mexican irregular. According to Lieutenant Smith, who was an eyewitness, "McClellan seized the Sergeant's musket, fired at, and killed the man who shot Hastings." The sergeant recovered to be commissioned as a second lieutenant at end of campaign.[35]

Smith and McClellan then reported at 10:00 P.M. to Worth, who "received us both very kindly," Smith later wrote, and "expressed satisfaction with the manner in which the works at the Garita had been carried." He also ordered the two lieutenants to "suspend operations for the night and resume them at daylight." Smith and McClellan, therefore, remained at the general's headquarters, which just after midnight received reports that Santa Anna had evacuated the capital and that civil authorities were suing for an armistice to surrender the city.

By 2:00 A.M., 14 September, the engineers were performing a reconnaissance to determine whether the report was true, with McClellan leading the detachment that examined the Alameda district in the heart of Mexico City. Having completed his mission, McClellan reported to Worth that indeed the Mexican army had evacuated the city, though Santa Anna had ordered the opening of the jails before making his escape, and the former inmates were carrying on a desultory resistance to the U.S. occupation. Until three o'clock in the afternoon, the sappers were engaged in street fighting, "particularly," reported Major Smith, "in breaking into houses with crow-bars and axes," killing a number of the Mexican irregulars, and capturing even more "suspicious persons."

For the greater part of the afternoon, McClellan commanded the company while Lieutenant Smith was searching for powder to be used in blowing up houses from which the Americans had been fired upon, "contrary to the usages of war." In one particular firefight McClellan reported killing more than twenty resistance fighters. In the week's fighting from 8 through 14 September, the engineer company had suffered two men dead and two wounded. Worth's official report on the capture of the city was fulsome in its praise of the engineer company, mentioning Smith and McClellan by name and inserting what he called "a respectful notice of the very intelligent enlisted men of the sappers and miners."[36]

1. Stevens Diary, 29 Mar. 1847, quoted in Stevens, *Life of Isaac Stevens*, 119.

2. Hillard, *Life and Campaigns of George B. McClellan*, 18–19.

3. McClellan to E. D. Townsend, 5 July 1861, in Sears, *Civil War Papers of George B. McClellan*, 45.

4. Hillard, *Life and Campaigns of George B. McClellan*, 18–20; Pillow to Scott, quoted in Randy W. Hackenburg, *Pennsylvania in the War with Mexico, the Volunteer Regiments* (Shippensburg, Pa.: White Mane, 1992), 323–24.

5. Scott to William L. Marcy, 23 Apr. 1847, in *Report of the Secretary of War,* 30th Cong., 1st sess., 1848, S. Exec. Doc. 1, 263; Stevens to Wife, quoted in Stevens, *Life of Isaac Stevens,* 127.

6. Stevens Diary, 20, 21 Apr. 1847, quoted in Stevens, *Life of Isaac Stevens,* 130–31.

7. Stevens Diary, 20 Apr. 1847, quoted in ibid., 130; Robert E. Lee to [?], 25 Apr. 1846, in Lee, *General Lee,* 40; Smith, *Company "A,"* 35; Stevens to wife, 1 May 1847, quoted in Stevens, *Life of Isaac Stevens,* 134–36.

8. Isaac Ingalls Stevens to Wife, 13 May 1847, quoted in Stevens, *Life of Isaac Stevens,* 140.

9. Hillard, *Life and Campaigns of George B. McClellan,* 19.

10. Ibid., 19–20.

11. Sears, *George B. McClellan,* 21; Hillard, *Life and Campaigns of George B. McClellan,* 19–20.

12. Isaac Ingalls Stevens to Wife, 14 May 1847, quoted in Stevens, *Life of Isaac Stevens,* 141–42.

13. Stevens Diary, 7–21 June 1847, quoted in ibid., 144, 148; Stevens to Wife, 27 May 1847, quoted in ibid., 144; Sears, *George B. McClellan,* 21.

14. Smith quoted in Sears, *George B. McClellan,* 21. Dr. George McClellan died on 9 May 1847 from what was described as "an ulcerative perforation of the small intestine." McClellan received the news in a letter from his maternal uncle, William White, the husband of Sara Frederica Brinton. William White to George B. McClellan, Philadelphia, 11 May 1847, George B. McClellan Papers, Library of Congress.

15. Stevens, *Campaigns of the Rio Grand and of Mexico,* 56.

16. Quoted in Stevens, *Life of Isaac Stevens,* 164.

17. Stevens, *Campaigns of the Rio Grande and of Mexico,* 58–59.

18. Beauregard, *With Beauregard in Mexico,* 42–43; Hillard, *Life and Campaigns of George B. McClellan,* 23.

19. Quoted in Stevens, *Life of Isaac Stevens,* 166–67; Hillard, *Life and Campaigns of George B. McClellan,* 23.

20. Stevens Diary, 14 Aug. 1847, quoted in Stevens, *Life of Isaac Stevens,* 167.

21. Smith, *Company "A,"* 39.

22. John L. Smith to Winfield Scott, 27 Aug. 1847, in *Report of the Secretary of War,* 30th Cong., 1st sess., 1847, H. Exec. Doc. 8, 349–50; Gustavus Woodson Smith to W. T. H. Brooks, 23 Aug. 1847, in ibid., 66.

23. Stevens, *Life of Isaac Stevens,* 171–172, 192 (quote); Smith, *Company "A,"* 40–41; Smith to Scott, 27 Aug. 1847, in *Report of the Secretary of War,* 30th Cong., 1st sess., 1847, H. Exec. Doc. 8, 352; G. W. Smith to W. T. H. Brooks, 23 Aug. 1847, in ibid., 67; "Report of General Twiggs," 23 Aug. 1847, in ibid., 322, 323, 324.

24. Quoted in Stevens, *Life of Isaac Stevens,* 170–72, 176.

25. Stevens, *Campaigns of the Rio Grande and Mexico,* 67.

26. G. W. Smith to W. T. H. Brooke, 23 Aug. 1847, in *Report of the Secretary of War,* 30th Cong., 1st sess., 1847, H. Exec. Doc. 8, 68, 72; Stevens, *Life of Isaac Stevens,* 180, 197; Stevens, *Campaigns of the Rio Grande and of Mexico,* 69.

27. Stevens to Wife, 22 Aug. 1847, quoted in Stevens, *Life of Isaac Stevens,* 196.

28. Beauregard, *With Beauregard in Mexico,* 55; *Report of the Secretary of War,* 30th Cong., 1st sess., 1847, H. Exec. Doc. 8, 353; *Report of the Secretary of War,* 30th Cong., 1st sess., 1847, H. Exec. Doc. 1, 248, 256, 263, 306, 310–15, 321, 323, 367, 385, 400, 403; Stevens to Wife, 22 Aug. 1847, quoted in Stevens, *Life of Isaac Stevens,* 199; Stevens Diary, quoted in ibid., 185.

29. Smith, *Company "A,"* 52; *Report of the Secretary of War,* 30th Cong., 1st sess., 1847, H. Exec. Doc. 8, 332–35.

30. Beauregard, *With Beauregard in Mexico,* 60.

31. *Journal of the Executive Proceedings of the Senate of the United States of America, 1845–1848,* 11 Aug. 1848; McClellan to Roger Jones, "Washington City, August 1848," vol. 1, McClellan Papers, Library of Congress.

32. Smith, *Company "A,"* 54.

33. Smith to Scott, 27 Aug. 1847, in *Report of the Secretary of War,* 30th Cong., 1st sess., 1847, H. Exec. Doc. 8, 428; Stevens Diary, quoted in Stevens, *Life of Isaac Stevens,* 212.

34. Stevens Diary, quoted in Stevens, *Life of Isaac Stevens,* 212–13; Smith to Scott, 27 Aug. 1847, in *Report of the Secretary of War,* 30th Cong., 1st sess., 1847, H. Exec. Doc. 8, 429; William J. Worth to Scott, 16 Sept. 1847, in *Report of the Secretary of War,* 30th Cong., 1st sess., 1848, S. Exec. Doc. 1, 392–93.

35. Smith, *Company "A,"* 63.

36. Ibid., 58–60; Smith to Scott, 27 Aug. 1847, in *Report of the Secretary of War,* 30th Cong., 1st sess., 1847, H. Exec. Doc. 8, 429–30; Worth to Scott, in *Report of the Secretary of War,* 16 Sept. 1847, 30th Cong., 1st sess., 1848, S. Exec. Doc. 1, 394.

Diary Entry

VERA CRUZ TO JALAPA

26 MARCH 1847–20 APRIL 1847

After the surrender of Vera Cruz we moved our encampment—first to the beach, then to a position on the plain between our batteries and the city. Foster was detached on duty with the other Engineers to survey the town and castle. Smith and myself were to superintend the landing of the Ponton and engineers trains, and to collect them at the Engineer Depot. Between the Quartermasters and Naval Officers this was hardly done when we left. I dismantled the batteries, magazines etc.—then amused myself until we left, with the chills and fever. James Stuart being too sick to go on with his regiment came over to our camp and stayed with us. Instead of being sent on in our proper position, at the head of Twiggs's Division, we were kept back and finally allowed to start on the same day that Worth started—we received no orders to move, merely a permission.

Our teams (6) were the worst I ever saw—they had just been lassoed as they swam ashore, and neither they nor their teamsters had ever seen a wagon before. We left Vera Cruz on the 13th [April]. By dint of applying some of the knowledge I had acquired under Guy Henry's parental care, I succeeded in getting four teams to Vergara (Twiggs's headquarters during the siege). As Smith and Foster did not come up I rode back to see what was the matter and found that they had arrived at a point opposite the middle of the city, broken down two sets of teams, got one teamster's arm and hand badly kicked—and the devil to pay in general. At last they got on, and by leaving half the loads by the roadside we managed by hard swearing to get to within one half mile of El Rio Medio by dark.

The road so far was horrible, being hilly and very sandy. Our mules were so weak and miserable that the men actually had to push the wagons along, and it was easy to see that our march was to be very severe upon all concerned. General Worth and his staff passed us as we were busily engaged in "cussing" a team up a hill—we then learned for the first time that Santa Anna was at Cerro Gordo with a large force. When we encamped this night everybody was tired to death, and the only event worthy of recollection was the thrashing that a certain lazy nigger "Isaac" received from his frisky "boss."

On the [14th] we made an early start and after "persuading" the mules up the hill beyond Rio Medio we got along without very much trouble until we arrived at Santa Fe. Here the wagons were unloaded and leaving me with about ten men Smith and Foster went back after the loads left at Vergara. Jimmie [Stuart] and I struck up an acquaintance with the Alcalde—a very nice sort of a man. I found a couple of cavalry barracks etc. We amused ourselves chatting with the Alcalde all day—who tried hard to stampede us with guerilla tales etc.

Captain Hughes[1] came up late in the afternoon, Smith arrived after dark, having left the wagons with the ordnance people about half a mile behind. While G. W. was at supper, Jimmy, who had been amusing himself by playing monte with the Rancheros, came back and amused us by an account of a muy poquito muchachito about four years old playing monte and smoking puros [i.e., cigars].

Foster came up at last, and we all turned in. Santa Fe is a poor little affair—no water, but rather a fine view of a large extent of rolling country.

On the 15th I started back after the wagons before daybreak "unwashed and uncombed." After a vast amount of swearing at "Seven Bottles," of whom more anon, I got all the wagons up to Santa Fe—set the men to work at loading the wagons—got my breakfast, and at last we started. Country at first a rolling prairie—finally more broken and woody. We passed some of the most magnificent forests I ever saw—trees covered with most beautiful flowers—the fields also—the villages were completely deserted. About the middle of the day we stopped at a stream to rest.

While taking our lunch under the bridge an old stupid Dutch teamster brought down his mules to water and finally proceeded to water himself. He drank seven (!) claret bottles full of water and at length finding that process too slow he took to his bucket! We went on and overtook the ordnance fellows at. . . . . . . Had a good supper and a fine sleep, although they did try to stampede us about Lancers etc.—but they could not do it.

Started early on the 16th [April]—country remarkably broken—even mountainous. We passed several very long hills, at which it was necessary to treble our poor little teams. Met Simon Buckner[2] with a beef party. Arrived at Puerto Nacional just before Worth's Division left it (about 2 P.M.). Saw all the fellows and made our preparations to start at twelve at night. Took a fine bath in the clear mountain stream, and then dinner. After dinner we went to see Santa Anna's Hacienda—found a little boy in it who was frightened to death at the Barbarians. A real [a coin worth about 12 cents] soon quieted him.

The bridge has a curved axis—it is a beautiful piece of architecture. It would be impossible to cross it were the heights around properly defended and the bridge itself occupied. The bridge and heights might all be turned by enterprising light infantry, for the stream is fordable. From the nature of the ground it would be impossible for artillery or cavalry to turn it without great trouble and labor.

Reveillé at 11:30—started at quarter past twelve—of course no undressing. Stuart "thought as he was already dressed there could be no hurry." Night pitch dark. About an hour before daybreak found in the road a saddle (American) and a pool of blood—some poor devil of a straggler from Worth's Division probably murdered. After ascending the hill just beyond this spot, G. W., J. S. and myself laid down in the road to sleep—that half hour's sleep just before going into battle was the sweetest I ever enjoyed.

Passed in the course of the morning a great many stragglers from Worth's Division—they had lagged behind in the night march. About two miles from Plan del Rio we were sitting in a rancho waiting for the wagons, when a wagon master came galloping by saying that the Lancers had cut off the train. The escort of dragoons was about 800 yards nearer Plan del Rio than we. We galloped back—the escort not far behind and found that our wagons were safe, but that the Lancers had cut off a few of the stragglers whom we had passed.

Suddenly a turn of the road displayed Plan del Rio at our feet—the little valley filled with troops, horses, artillery, wagons, etc. We arrived at about 10.30 A.M.—found the Engineers and took a lunch with them. G. W. S. and myself then rode out to Twiggs's position with Captain Lee—we arrived just in time to see the ball open. Saw old Twiggs, who wondered "Where the devil did you two boys come from?" and started back to bring up the company. On the way back a round shot came about as near my head as would be regarded agreeable in civil life and then missed enfilading the 2nd Infantry about a foot and a half.

When we got back to El Plan, I was ordered to join Tower with ten men—to go with Gid Pillow and the Mohawks. Did my best that afternoon to find out where we were to go in the morning but none of them would tell me anything about it. G. W. left me ten of the best men in the company, and took Foster and the rest with him to report to General Twiggs. It seemed to be a mutual thought that the chances all were that we would not meet again! The idea of being killed by or among a parcel of Volunteers was anything but pleasant.

Got up before daybreak—woke up the men—had the mare fed and saddled—drank some coffee—distributed tools to my party and was ready for battle long before our dear Mohawks had their breakfasts. Also gave some tools to the Volunteers. My men had hatchets, axes and billhooks—the Volunteers axes, sap-forks and billhooks. At length all was ready and much to my surprise we marched straight up the road toward Jalapa.

So little did I know of our point of attack—I only knew that we were to attack either their right or front, and that we would as surely be whipped—for it was a Volunteer Brigade. I led off with my detachment, and after passing the greater part of Worth's Division—which was formed in column of platoons in the road—we turned off to the left, nearly opposite the point where Twiggs turned to the right. Tower directed me to place my men on the path inclining most to the left. I did so and rested my men, whilst waiting for the Volunteers who were a long distance behind. At length General Pillow came up, and seeing my men, directed that they should be placed on the path inclining to the right.

Lieutenant Tower made some remark about changing the route, and also that we would be more apt to be seen when crossing some ravine if we went to the right. I remember distinctly that the impression made upon me by the conversation was that General Pillow had against the opinion of Lieutenant Tower changed the route to be followed in order to attain the point of attack. I had no idea of the importance of the change and that it could lead to a different point of attack. I afterward found that the different paths led to very different parts of the enemy's position, the one we actually followed bringing us in a very exposed manner against the front of the works, whilst if we had taken the one advised by Lieutenant Tower we should have turned the right of their works and have been but little exposed to their fire.

The fault of the erroneous selection was General Pillow's, except that Lieutenant Tower should, as the senior Engineer with the column, have taken a firm stand and have forced General Pillow to have pursued the proper path. It was certainly a fine opportunity for him to show what stuff he was made of—but unfortunately he did not take advantage of it at all.[3]

We at length moved off by the flank. My detachment at the head, and during the movement—at all events before the firing against us commenced—we heard the musketry of the attack of Twiggs's Division upon the Telegraph Hill.

After moving about two-thirds of a mile from the main road we reached a certain crest bordering upon a ravine, whence a strong picket of Mexicans was observed. Tower advised General Pillow to incline his Brigade well to the right in order to cross the ravine lower down and out of view. The General directed Colonel Wynkoop[4] to countermarch—file twice to the right and move upon a certain dead tree as his point of direction (Colonel Campbell's Tennessee Regiment[5] to support him). He was then to form his men for the attack and charge upon hearing a concerted signal from the rest of the Brigade.

Colonel Haskell[6] at once commenced forming his Regiment in a column of platoons, the flank of the column toward the work. His men having straggled a great deal, this arrangement was attended with some difficulty—the men being literally shoved into their places one by one. Hardly two platoons were formed when General Pillow shouted out at the top of his voice—"Why the H—l don't Colonel Wynkoop file to the right?" I may here observe that we had heard very distinctly the commands of the Mexican officers in their works. This yell of the General's was at once followed by the blast of a Mexican bugle and within three minutes after that their fire opened upon us. The General may have shouted this before a single platoon of Haskell's was formed—but the interval must have been very short, because Wynkoop's Regiment had not reached its destination and had not formed there when the firing commenced.

When the Mexican fire opened, Haskell's Regiment became at once "confusion worse confounded."[7] Some of the men rushed toward the works, many broke to the rear, very many immediately took cover behind the rocks, etc. I at once asked General Pillow for orders to proceed "somewhere" with my detachment—for I had as yet received no orders or directions from anyone and was utterly ignorant of the ground.

While talking with the General—who was squatting down with his back to the work—he was wounded in the arm, upon which his aide, Lieutenant Rains,[8] appeared from somewhere in the vicinity and they together went off to the rear, on the run. I then went in amongst the Tennesseeans and found at once that it was useless to attempt doing anything there, as that Regiment (Haskell's) was utterly broken and dispersed and the Pennsylvania Regiment, which was to support them, had kept so well in reserve that they could not be found. I then went over to the other side of the ravine—the firing had by this time nearly if not altogether ceased.

Upon arriving there I found Campbell's Regiment in pretty good order and in good spirits, the Pennsylvania Regiment (Wynkoop's) in most horrible confusion. Campbell was moving on toward the work, and I at once advised General Pillow to halt him until some order could he restored to the other Regiments. He took my advice and directed me to give the order to Campbell, which I did.

I thought that it was by no means certain that Campbell alone could carry the works and that if he were checked or repulsed all was lost, for there was not a company formed to support him. Besides, although his Regiment was moving on well, they were not then under fire, nor had they been under any fire, to speak of, that day—so I doubted the steadiness of their movements when their advance should have brought them in sight and under the fire of, the Mexicans.

Colonel Haskell came up without his cap about this time and a very warm conversation ensued between him and General Pillow—the General accusing him of misconduct and deserting his troops, the Colonel repelling his assertions and stating that his Regiment was cut to pieces. I at once, without saying a word to either the General or the Colonel, called to my party and directed them to beat the bushes for "2nd Tennesseeans" and to bring all they could find to where we were. They soon returned with quite a number.

In the course of conversation I told General Pillow that I did not think that he could carry the works without some Regulars. He assented and directed me to go at once in search of General Scott and ask him, from him (Pillow), for a detachment of Regulars—whatever number he could spare, saying that he would make no movement until my return.

I immediately ran down to the road where I expected to find General Scott and Worth's Division and there found that the General had gone on. I jumped on my mare and galloped around by Twiggs's road and at length found the General about half way up the ridge over which Worth's Division passed to reach the Jalapa road—the rear of Worth's Division was then crossing. I told the General my message and he directed me to say to General Pillow that he had no Regulars to spare, that the last of Worth's Division was then passing over, that Santa Anna had fallen back with all his army, except about 5000 men, toward Jalapa, that he expected to fight another battle with Santa Anna at once, and that he thought it probable that the 5000 men cut off would surrender—finally that General Pillow might attack again, or not, just as he pleased. He evidently was not much surprised and not much "put out" that Pillow was thrashed, and attached no importance to his future movements.

With this reply I returned and could not for a long time find any of the valiant Brigade. I at length found Wynkoop's Regiment. He told me that white flags were flying on the work and that one or two had come down toward his position—but that as he did not know what they meant, could not raise a white handkerchief in the crowd, and had no one who could speak Spanish, he had held no communication with them.

I told him what they meant and said that when I had seen General Pillow I would return and go to meet them. As I left he asked me if I could not give him an order to charge—I said "No"—then said he—"Tell General Pillow that if I don't get an order to charge in half an hour, I'll be d—d if I don't charge anyhow"—this after I had told him that the white flag meant a surrender!!!

I at length found General Pillow some distance in rear and reported. Castor[9] came up a moment or two afterward and told General Pillow that he had been sent to inform him that the Mexicans had surrendered—on which I took my men down the road and directing them to come on and rejoin the company as soon as possible—I galloped on to overtake it.

During my conversation with General Scott he mentioned that he had seen the charge of Twiggs's Division and spoke of it as the most beautiful sight that he had ever witnessed. He said everything in praise of his "rascally Regulars."

With reference to the operations of Twiggs's Division.—During the afternoon of the 17th [April] the hill opposite to and commanded by the Telegraph Hill was carried by Harney's[10] (Smith's) Brigade and the enemy pursued partly up the Telegraph Hill by the Rifles and 1st Artillery. They were, however, recalled to the hill first mentioned, which was occupied in force.

During the night one twenty-four-pounder, one twelve-pounder and a twenty-four pound howitzer were with great difficulty hauled up and put in position behind a slight epaulment. There were also a couple of the Mountain Howitzers and some Rocketeers. Shields's[11] Brigade of Volunteers were somewhere in the vicinity to support and were employed to man the drag ropes used to haul up the pieces. It may be well to mention that they were more than once "stampeded" while engaged in this by the mere discharge of a piece—no ball coming near them.

Another detachment of New York Volunteers was engaged during the afternoon and night of the 17th in hauling an eight inch howitzer along the crest on the other side of the "Rio" in order to take an enfilade or reverse fire upon the Mexican works. Taylor's Battery was with Twiggs, Duncan came around with Worth—Steptoe[12] was with Twiggs. The cavalry and rest of the artillery were in the Jalapa road ready to advance in pursuit.

Harney was directed to storm the hill, Reilly[13] to cut off the retreat of the Mexicans by the Jalapa road—Worth to support. The affair of the 18th was opened, on our side, by the fire of our artillery. The 24-pounder was badly served and did little or no real damage. At length Harney charged over the valley with the 1st Artillery, 3rd and 7th Infantry, the Rifles being thrown out to cover his left. He carried the hill in gallant style.

Riley allowed himself to deviate from his proper path and instead of pushing straight on for the Jalapa road, he amused himself by skirmishing to his right and left—so that he did not accomplish the purpose for which he was sent, that is, he did not cut off Santa Anna's retreat.

In the meantime Shields was sent around still further to our right, to turn the Mexican left. He finally came out in front of certain batteries, charged them but was repulsed completely and himself badly wounded. About this time Harney carried the Telegraph Hill and that commanding these last batteries, one or two discharges from its summit with the captured pieces at once cleared them. Upon that the Volunteers right gallantly charged and carried them at the point of the bayonet, there not being a soul in the battery at this time.

Twiggs—at least a part of his Division—moved on at once in pursuit. The Cavalry soon followed, but the Mexicans had gained a long start and made the best use of their legs—so that not very many were killed or taken in the pursuit. Twiggs and the Cavalry, also the Volunteers, halted at Encero. Worth remained at Plan del Rio and Cerro Gordo. I myself overtook my company at Encero where we bivouacked that night—and felt right proud that we had won that day a glorious victory.

On the morning of the 19th we marched from Encero to Jalapa, about twelve miles, at the head of Twiggs's Division. We entered Jalapa about 11:30 A.M., our company being the first American infantry to set foot in that city. It rained quite violently during the greater part of the march, which prevented me from enjoying fully the beauty of the scenery, especially as I had to foot it.

It was really delightful, upon entering Jalapa, to see gentlemen and ladies, at least persons dressed and appearing as such. The white faces of the ladies struck us as being exceedingly beautiful—they formed so pleasing a contrast to the black and brown complexions of the Indians and Negroes who had for so long been the only human beings to greet our sight. The Jalapeños appeared perfectly indifferent about us, manifesting neither pleasure nor sorrow at our approach. Our march from Encero and entrance into Jalapa was entirely undisturbed—not a shot being fired or soldiers seen. Of course not the slightest excess was committed by any of the Regulars. We at first marched to the Cuartel where we remained some few hours, until at last we were ordered to a posada on the Plaza.

I was very much pleased with the appearance of Jalapa and its inhabitants. The women were generally pretty, the gentlemen well dressed. They carried to a great extent the custom of filling the balconies with flowers, which gave a very pleasant appearance to the streets.

Soon after we had established ourselves at the posada we were astonished by a great commotion in the streets, which was ascertained to be caused by the arrival of the Cerro Gordo prisoners who had all been released on parole and of course fought us again upon the first opportunity. They were marching back to Puebla and Mexico, organized in regiments, etc.—merely being deprived of their arms. The disgust in the Division at this release was most intense, we felt poorly repaid for our exertions by the release of these scoundrels, who, we felt sure, would to a man break their parole. They passed the night in the streets around the Plaza and in the morning robbed all the poor market women in the vicinity.

We had no beds that night—our baggage not being up—were lucky enough to get some frijoles and chocolate for supper—breakfast ditto. Worth's Division came up about one o'clock on the 20th and we were ordered on at the head of it,—to leave Jalapa at 3:30 of the same day.

1. George Wurtz Hughes attended West Point from 1823 through 1827 but did not graduate. In 1838, however, he was commissioned a captain of topographical engineers. He was chief engineer on John E. Wool's staff in 1846 and on William Jenkins Worth's staff in 1847. From December 1847 until the evacuation of Mexico in 1848, he served as civil and military governor of the Department of Jalapa and Perote.

2. Simon Bolivar Buckner, class of 1844, was brevetted to first lieutenant in the Sixth Infantry for service in the Battle of Churubusco, during which he was wounded, and to

captain at Molino del Rey. Following the war he returned to the U.S. Military Academy as assistant instructor of infantry tactics.

3. Isaac Stevens disagreed. In his estimation Tower, "for judgment, for an assured and natural self-reliance, great force of character, and great decision and intrepidity in emergencies, has no superior in our corps." Stevens, *Life of Isaac Stevens,* 217.

4. Francis Murray Wynkoop was elected colonel of the First Pennsylvania Volunteers. After the war Franklin Pierce appointed him marshal of the Eastern District of Pennsylvania.

5. William Bowen Campbell, U.S. congressman and Seminole War veteran, was elected colonel of the First Tennessee Volunteers in General Pillow's brigade.

6. William T. Haskell was colonel of the Second Tennessee Volunteers. All primary accounts of the Battle of Cerro Gordo agree that Haskell's attack on the Mexican batteries was a grave and costly blunder. According to the *Niles Register* of 24 July 1847, of the 1,040 officers and men of this regiment, only 360 survived the campaign, the rest succumbing to "sickness and bullets, disease, and shot and swords."

7. McClellan alludes here to book 2, lines 995–96, of John Milton's *Paradise Lost:* "With ruin upon ruin, rout on rout, Confusion worse confounded."

8. George Washington Rains, class of 1842, was commissioned into the engineers but transferred to the Fourth Artillery in 1843. He received brevets to captain and to major for his conduct during the battles for Mexico City.

9. Thomas Foster Castor was graduated in McClellan's class of 1846 and commissioned into the First Dragoons.

10. William Selby Harney was the senior cavalry officer under Winfield Scott. The general, however, mistrusted Harney's judgment and relieved him of command. President Polk, a neighbor of the Harney family in Tennessee, overruled Scott's decision, but the ensuing friction was largely extinguished by Harney's conduct during the Battle of Cerro Gordo. He was brevetted to brigadier general and promoted to the permanent rank of colonel of the Second Dragoons on 30 January 1848.

11. James Shields, one of several political generals appointed by the Polk administration, received command of the Illinois brigade at the rank of brigadier general and was brevetted to major general following the Battle of Cerro Gordo. He thereafter was placed in command of a brigade of New York and South Carolina volunteers in Patterson's division, which he led with no particular distinction for the duration of the Mexico City Campaign. Zachary Taylor described Shields as "without one particle of principle to restrain him, save the laws of his country and ready to minister body and soul to the vilest passions of a vile administration." Quoted in Smith, *War with Mexico,* 1:352.

12. Edward Jenner Steptoe, class of 1837, was brevetted to major and then to lieutenant colonel in the Third Artillery for meritorious conduct during the battles of Cerro Gordo and Chapultepec.

13. Bennet Riley, a veteran of the War of 1812, commanded a brigade of Twiggs's division in Mexico. He was brevetted to brigadier general and to major general for his actions at Cerro Gordo and Contreras.

George B. McClellan to Frederica M. English

CASTLE OF PEROTE, MEXICO

22 APRIL 1847

I have only time to tell you, my dear Sister, that I have got through the battle of Cerro Gordo safe and sound. It was a most beautiful affair and followed by great results. The Mexicans were so much scared by their unexampled rout that they evacuated Jalapa, Passeo de la Hoya and this beautiful Castle, and we have entered all without firing a gun. It is even said that they have evacuated Puebla. We leave for that place in about two days. Worth's Division is alone here. Twiggs and the Volunteers are at Jalapa.

Jalapa is the most beautiful place I ever beheld and the surrounding country is a perfect Paradise. At Vera Cruz we were roasted to death; here the same end is very nearly effected by freezing. I just saw in the chapel of the castle the tomb and coffin of Guadalupe Victoria, the greatest man whom Mexico has produced and her first President.

We have been on the move ever since the fight. We advanced ten miles on the day of the battle after thrashing them. You will think this [at] best a poor apology for a letter, but the mail leaves in a few minutes, and it is in all probability the only one that will leave before we start for Puebla, so it is this or nothing. Give my love to Tom and the children and believe me to be

Your affectionate brother,

Geo. B. McClellan

Songo, Jim, and the mare are in splendid condition. Jim is as fat as a buck and as impudent as possible.

6

"NO ONE CAN SAY 'POOR MAC' OVER ME"

MEXICO CITY

24 October 1847–6 June 1848

Although the fall of the Mexican capital did not in itself end the war—negotiations would drag on for nearly five months until the Treaty of Guadalupe Hidalgo was signed—the 13 September battle effectively ended the fighting. On the evening of the fourteenth, when Smith and McClellan had returned to their quarters after experiencing the emotional denouement of a long and arduous campaign, Smith noticed that his comrade was "very quiet for a considerable time, evidently thinking of matters that deeply interested him. An occasional marked change seemed to come over the spirit of his dream." After some time Smith "awakened him from his reverie, saying: 'A penny for your thoughts. I have been watching you for half an hour or more, and would like much to know, honor bright, what you have been thinking about.'"

"I have been making a 'general review' of what we have gone through since we left West Point," McClellan replied, "one year ago this month, bound for the 'Halls of the Montezuma'; have been again on the Rio Grande, that grave-yard of our forces, have gone over the road from Matamoras to Victoria and Tampico, where we had so much hard work; went through the siege of Vera Cruz, where we were left out in the cold during the ceremonies of surrender, and later, had to make our way as best we could, with the engineer train through the horrid sand, glanced at Cerro Gordo, where it was my misfortune to be with General Pillow's 'whipped community'; stopped again with our friends, the Monks, in the convent at Puebla, crossed over the mountains; came by way of San Antonio, Contreras, Churubusco, Chapultepec and the San Cosme Garita, into this city. Here we are—the deed is done—I am glad no one can say 'poor Mac' over me."[1]

To the contrary, William J. Worth was liberal in his praise of his engineers during the battles in and around the Mexican capital, naming particularly Stevens, Smith, and McClellan for having "displayed the gallantry, skill, and conduct, which so eminently distinguished their corps." And although he spent eight long months in Mexico City acting as company quartermaster, McClellan often went sightseeing, attended the opera, and visited the Aztec Club, where, he wrote home, "We will meet nothing but gentlemen." Moreover, the quarters he shared with Smith were in the palatial home of Gen. of Division Manuel María Lombardini. But McClellan yearned for home.[2]

Shortly before the Mexican negotiators agreed to terms on 2 February 1848, the army's senior engineer officer, Robert E. Lee, granted Smith's request for permission to turn over command of the company of engineers to McClellan and return to the United States. On 28 May George McClellan led the sappers and miners out of Mexico City for embarkation at Vera Cruz. From there they were transported by steamer to New York, arriving on 22 June, and thence to West Point, where McClellan turned over his command to Capt. George W. Cullum. Of the seventy-one men of the company who had left New York in September 1846, two had been killed and six wounded.[3]

Although himself eager to return home, when his orders at last came, McClellan was grieved to leave the military community that had been his world for the past eighteen months. "The attachments formed in a campaign with men by whose side you have fought," he wrote to his mother, "with whom you have exulted in the moment of victory and whom you have met day after day—such attachments cannot be severed without a pang." Lamenting the end of an active campaign, but little knowing the irony of his prophecy, on the eve of his departure for the United States, McClellan resolved to "pass the rest of my life in a very tame and humdrum manner."[4]

1. Smith, *"Company 'A,"* 64.

2. *Report of the Secretary of War,* 30th Cong., 1st sess., 1847, H. Exec. Doc. 1, 385; Smith, *"Company 'A,"* 63; McClellan to Elizabeth B. McClellan, 24 Oct. 1847, George B. McClellan Papers, Library of Congress; *Journal of the Executive Proceedings of the Senate of the United States of America, 1845–1848,* Apr. 12, 1848.

3. Smith, *Company "A,"* 72–73; Joseph G. Totten to William L. Marcy, "Report of the Chief Engineer," in *Report of the Secretary of War,* 31st Cong., 1st sess., 1849, H. Exec. Doc. 1, 281.

4. McClellan to Elizabeth B. McClellan, 22 Mar. 1848, McClellan Papers.

George B. McClellan to John H. B. McClellan

CITY OF MEXICO
24 OCTOBER 1847

I have at last, my dear John, an opportunity of writing a decent sort of a letter to you since the first trains that have left this city will start in two or three days for Vera Cruz. We have received no mail from the U.S. since sometime in July, and I got nothing by that except one letter written by our Father some little time before his death.

Of that event I do not and can not speak, for it can but call to our minds the remembrance of as noble a being as graced the earth, and I know that the feelings of both of us are too sad to make it proper to allude to it unnecessarily. I hope and trust that time has softened the severity of the blow to our dearest Mother, and I have purposely forborne making any allusion to that subject in any of my letters to her after the first one I wrote about it.

So I am by no means confident that you have received any of the letters I have written to you on the subject. I will repeat my wishes concerning Father's estate, & c. I am well aware that by being yet a minor, I have no legal voice in the matter, but I am equally certain that you will wish to know my opinions and wishes in the matter, and that you will be guided by them exactly as if I were of age and on the spot. In the first place, I wish the family to remain in the old house if it is possible.[1]

In the next place, if after the debts of the estate are paid there is any thing left, I wish all, every cent, of my share to be made over to Mother for her use and that of the two children. If it is necessary that I should sign any legal form in order to effect this arrangement, send it to me at once by the mail and I will sign and send it back by the very first chance.

In the meantime, act as the laws require, but use my share for Mother. I must insist upon it that you [not] act otherwise and that you use for the present at least your share. My reasons for this are that in the first place I have a certain, though moderate, income, ample for my present wants, and I have no idea of changing my condition, that is of marrying; whereas with you the case is different. You have been practicing but a short time, and your income can be but moderate; you have a certain position to occupy and in all probability your present income is not sufficient to suit it.

I know that you will shortly marry, in which case your necessity will be greater, and I therefore insist upon it that you act as I have requested. I am equally certain that many years will not elapse before your Profession will render you independent of all other sources of income, for I feel confident that you must before long occupy the first place in your Profession, and when that day comes you can, if it is necessary, then give to our Mother your share of the estate.

Do write to me and tell me at some length the whole state of affairs, for it is of great importance that I should know to what extent the family is dependent upon me for by curtailing my extra expenses. I can doubtless without inconvenience be of some assistance to you.

I send you with this a check for $400.00 for the propose of paying my debts in the U.S. which are as follows: To John Smith Fraser, Broadway, N.Y., opposite City Hotel and over St. John's Hat Store: $183.00 (one hundred eighty three dollars), for 1 pr. epaulettes, $50.00; 1 full dress coat, $55.00; 1 undress frock coat, $30.00; 1 rapier, $25.00; 1 jacket, $15.00; 1 castle [i.e., the emblem of the Corps of Engineers], about, $2.00; say, $180.00.

To Wm. T. Jennings' and Co., Tailors, 231 Broadway, N.Y. (next to American [Hotel]), $138.38, but I sent him a check for $82.83, so that I now owe him $55.55 (fifty-five dollars, fifty-five cents).

To John Earle, Jr., and Co., Tailors, Washington and Boston, I owe $70.50 (seventy dollars and fifty cents). I wish you, then, to send $180.00 to J. Smith Fraser, $60.00 to Jennings, $75 on to Earle, in all, $315.00. You will find Jennings' and Earle's bills in a drawer in a dressing case in the back 3rd story room in with a number of my papers & c. But I have got other things of Jennings since I received that bill and you will observe that I have marked certain articles on Earle's bill as not having been received.

Get receipts for me from them all, in full if you can, and from Jennings also get a receipt for the check, $82.83, I sent from Vera Cruz in April or March. The balance of the check, $85.00, be kind enough to accept from me and use it for mother or yourself as circumstances may require.

Attend to this matter for me at once, and keep the receipts, merely letting me know that they are settled. If you should be going on to Boston soon after receiving this, attend to it in person if not paid by letter. A great load will be off my shoulders when these debts are paid, for I am then straight, square, and fair in the world. I will rather not owe anyone a cent, and can assist Mother if she is at all in want of it.

Do tell me plainly, frankly, and fully the exact condition of affairs at once. I must know them at some time or other, and I should never forgive myself if any of you had been suffering in the slightest when I could have remedied it.

Enough of this business for now. I must know that you understand me, and that when I receive a reply we will understand each other fully. Let me beg you to tell mother, put the best appearance on everything; look at the bright side of affairs and trust me that everything will, or will in the end, it shall be so if I can do anything towards it.

I feel so glad and proud that I have got safely through the battles in the war that it will take a heavy, heavy shock to make me despond. Thank God our name has not suffered so far in my hands. Well! Well! I will be truly glad when this war is over and we leave this county. I am tired of it, heartily so; so are all of us; yet it is impossible even to guess when the wretched thing will end. I should not wonder if I staid here five years nor be astonished to get off in a month. Nothing can surprise me now. I have been drilled to wonders since I left the states.

Tuesday—well, I must finish this scrawl and mail it, for the train will not wait for me. Stuart and G. W. Smith send you and all of you their fondest regards. Smith had some hopes of getting out of the country in the train in order to recruit the Company (we only have about 30 men), but he was unable to effect it. I would in that case have been left in command of the Company.

We will soon hear, I hope, something definite about the permanent command of our poor little forlorn hope—whether it is to remain with G. W. or be given to some "old fogy" of a Captain. I hope that they will give it to G. W. and let us all go out to fill up the Company. We have now too few men to effect any very great object. We will probably have more companies authorized at this session of Congress.

Captain Mason of my Corps, wounded at Molino, goes out in this train. He will take my letters to New Orleans and mail them. As his wife is in Newport, he will probably pass through Philadelphia. If you hear of him at the hotel, do me the favor to call on him. You will find him a splendid fellow. Do write soon to your affectionate brother,
Geo. B. McClellan

I have a pretty little *puro* [cigar] case with the Mexican arms on it, which I bought at Puebla for you, but I am afraid that it is almost too large to go in a letter so I will keep it till I come myself. How are the filly and the dog? If it is in the slightest degree necessary, do not hesitate to sell my horse, but do not let Father's watch go. I will make any sacrifice, buy it at any price, rather than that should happen. My very best respects to the Rawles. Have you left your old office, over the way? How comes on the Club?

1. The McClellan home was located at 248 Walnut Street, at the corner of Seventh Street, now in the heart of downtown Philadelphia.

George B. McClellan to Elizabeth Brinton McClellan

MEXICO CITY

24 OCTOBER 1847

At last my dear Mother a train is going to Vera Cruz, and a mail with it so that I feel quite certain that the letter, at all events, will go through safely. All those that we have sent heretofore were by such conveyances that we could not be certain that you would receive our letters, and even such opportunities were few.

It is now utterly impossible to form any idea of the length of time we will have to remain in this country, for there is now no government in the country but that of our army, and no one can guess what will be the policy of our government in future. I trust, tho, that you will feel contented that I have gone through these battles without wounds and, thank God! without discredit, and that the thoughts of this may reconcile you to my being in this country for a long time.

We are very pleasantly situated, indeed; we have a very good house to ourselves (G. W. and myself, with the Company directly on the Alameda). We live on the fat of the land, have just enough to do to make time pass pleasantly. We have, too, the magnificent Theatre of Santa Anna open every night, besides there has been formed amongst us of the old Army, a club (like John's) called the "Aztec Club."[1] We have a magnificent club house, and it is a source of great pleasure and comfort to us. We go there and are sure that we will meet none but gentlemen.

We have, of course, but little information in reference to the remains of the Mexican Army. It appears, however, to be quite certain that they cannot collect in sufficient numbers to annoy us. We hold the fountainhead of their money and supplies by being in this City. You may rest assured that they will never come here to attack us. They have been thrashed too badly in this valley to do that.

I see some very fine views of the City and of different places in Mexico which I will buy before I leave. I think I should be wise to follow Capt. Lee's advice, that is, to write to my sweetheart to take somebody else as I shall not return for at least ten years. However, "nous verrons," I really can't see any reason why we should get home in ten years rather than ten months. I think we will have seen quite enough of Mexico long before we leave it.

I would greatly prefer starting on a march again. It is a most delightful way of spending time, I assure you. If we must stay here, I should rather spend my time in traveling over the different parts of the country than to remain all the time in this City. I have, of course, been to see the Museum, which is a great humbug. There are some few objects of interest such as the old sacrificial stone, idols, arms, &c. of the Aztecs, but as I am now an Aztec myself, they are but everyday matters to me. The old Sacrificial Stone is a cylinder about four and one half feet high and eight feet in diameter. It is covered with carvings representing combats &c., quite roughly done. In the centre of the top is a basin and a trough to receive and carry off the blood of the victims.

It does make one feel a little strange to stand on the very block where have been immolated thousands of victims to the fierce old Indian Deity. Over the very spot where stood Cortes and his brave little band, listening to the thrilling yells of agony uttered by their comrades who were so unhappy as to glut the maw of the old War God before their very eyes, now floats one brave old banner [the "Stars and Stripes"], on the very spot where Montezuma's feather flag was displayed over his palace and where the royal banner of Spain was displayed over the dwelling of her viceroys, and where (to use Billy Jenkins', alias Worth, remarks) "the gaudy flag of the Mexicans" proudly waved defiance to us.

I bought the other day a pair of silk mittens for Mary, which I will enclose in this letter. Tell her that they come from a store on the Grand Plaza opposite to the Palace called "Cajon al Tocador de las Damas"—"Store to the Toilet of the Ladies"—and if they are not as handsome as I could wish, it is because I could not send embroidered ones in a letter. She must wear them as a Mexican trophy.

As for Mr. Arthur, I don't think I can find anything to send him in a letter. I will look, however. I have the two rag figurines which I bought in Puebla, representing an old Mexican woman and a man going to market. They are very handsome and very natural. If I can send them by Major Smith, I shall do so. They are for Arthur and Mary together, and I shan't like it if they give away anything I send them for themselves. These figures will make a handsome ornament for the parlor. Don't let them be pulled about and destroyed.

The Italian Opera Troupe is to bring out *La Sonnambula*[2] tomorrow night, with magnificent scenery &c. The Spanish company has been playing this week in a play called the Magician of Seville. The scenery and dresses are the most magnificent I ever saw. The Theatre itself is said to be second but to that of San Carlos at Naples, and indeed it fully sustains its reputation. It is very large—probably twice as large as the Walnut.[3] It has five tiers of boxes and is superbly decorated. The ladies are beginning to frequent, especially on Sunday evenings.

There is also an American Company playing here, but it is rather too bad to be endured. You would have been amused when we first came here to have heard the "Bhoys" calling for "Yankee Doodle," "Star-Spangled Banner," &c. in the "Gran Teatro de Santa Anna," not a week before graced by his presence and that of all his generals. One would have thought that we had thrashed them!

Tuesday afternoon—The train goes on Thursday morning, and letters must be mailed tomorrow. The Sonnambula was produced last evening by the Italian troupe. They did very well, though not as well as the Seguins, &c.[4] Tonight the Spanish Company. In order that you may not think that we live like Barbarians exactly, I'll tell you that "ourselves" (G. W. and myself) have fine tablecloths, china plates, cups, &c. (gilt, at that,) excellent knives and *silver* forks. (i.e., Britannia).[5] We have omelet every morning for breakfast, soup, ducks, snipe, &c., &c. for dinner. We live like "fighting cocks."

As I can find nothing else for Arthur that will go in a letter, I will send him a "Quartilla," which he can either keep as a curiosity or expend in gingerbread, candy, &c. If he adopts the latter alternative, I have merely to request that he does not make himself sick, nor allow himself to plunge into extravagant habits on the strength of it. He must remember that *Old Johnny Swift* is a terrible fellow on all those who do not pay their debts, e.g., *Arthur and his penny owed to Skagarach, Pattegal, &c.*

Another train is to leave in six weeks. You must not expect to hear from me before that time. Do write often, my dear Mother, and remember that the ordinary mail is the safest conveyance.

Yours affectionately,
George

[postscript, in margins:] Remember me in the kindest manner to Aunt Steinmetz—indeed, to all.

I understand that General Scott has hope of a peace some time next spring. I do not know on what he based his opinion, but do not believe that we shall be so fortunate as that. The General is very kind to me when I see him and always blows me up for not coming to dine with him. I never go except when sent for on duty. I prefer keeping clear of Generals.

Hurrah! Ain't I glad I'm alive. I tell you what, one can't tell what fine fun it is to live until he has been through about six battles!

They have some very fine engravings in this City representing the City itself and many others in the self-styled "Republic."

Ask Arthur if he don't want to come out here and relieve me. I'll give him my Mexican blanket if he will. Tell Mary that I expect she'll be able to walk and talk pretty plain English when I get back.

1. The Aztec Club was founded on 13 October 1847 with its membership drawn from the officers of both the regular and volunteer regiments of the U.S. Army. In the words of Bvt. 1st Lt. DeLancey Floyd-Jones of the Fourth Infantry, "the Club was organized for the purpose of forming a resort for officers, as a promoter of good fellowship, and of furnishing a home where they could pass their leisure hours in social intercourse, and where more palatable and healthful viands could be procured at a reduced price than at the best Fondas of the city." The club was housed in the former residence of Mexico's former minister to the United States, Boca Negra, on the Plaza de la Constitución. John A. Quitman was the organization's founding president.

2. Vincenzo Bellini's *La Sonnambula,* or *The Sleepwalker,* an opera in two acts, was first performed at the Teatro Carcano in Milan on 6 March 1831.

3. The Walnut Street Theatre is the oldest theater in the United States, operating continuously since 1809. Every noteworthy American actor of the nineteenth century is said to have performed there.

4. Soprano Anne Childe Seguin and her husband, bass Edward Seguin, were the founders of the earliest English opera troupe to make a permanent home and career in the United States, relocating from England in 1840. For the next ten years, they dominated the performance of opera in English in America.

5. Britannia standard, an alloy containing 95.84 percent silver, was introduced in England by an act of Parliament in 1697 in an effort to limit the melting of sterling-silver coinage. Its hallmark is "the figure of a woman commonly called Britannia."

George B. McClellan to Sen. Daniel Sturgeon[?]

CITY OF MEXICO

30 OCTOBER 1847[1]

My dear Sir:

Owing to the uncertainty of the mails which have been sent from here to Vera Cruz (by private couriers &c.) I fear that a letter which I took the liberty of writing to you some few weeks ago has never reached you, as it was upon a subject of deep interest to me, in common with all the younger officers of the Army. I am induced to trouble you again upon the same matter.

It referred to the appointments of citizens in the Army over the heads of the Brevets already in service, who had been fortunate enough to have served during the latter part of the Rio Grande campaign and during the whole of this. I mentioned as a glaring instance, indeed the very object of my writing the letter, the appointment of one Christopher Carson,[2] a protégé, I believe, of our would be General in Chief, Lieutenant General and Minister Plenipotentiary [Thomas Hart Benton]—as a 2nd Lt. of Mounted Riflemen over the heads of all those of my own class who had been in the Regiment for more than a year and who have the honor of having contributed to no small degree to the reputation so nobly earned by that gallant Regiment.[3]

If you have not received my letter, you will have learned before this time by the official reports that two of them have very highly distinguished themselves in the late actions—one (Stuart) was the first man in every battery between Tacubaya and Mexico and the very first man of the American Army in the City of Mexico. I will not insult you, sir, by asking you whether it is your opinion that Mr. Kit Carson or, Sir, anybody else should be appointed over such men as these. If such is to be the case, what inducement have we, in heaven's name, to exert ourselves in the slightest degree? You cannot imagine, Sir, the effect which has been produced upon the Army (the old Regulars, I mean) by the course adopted by our executive!

In time of peace, we were consoled for our slow promotion by being told that when war came the Army would be increased and that *then* we would have rapid promotion, that the higher positions of Field Officers and Captains would be filled from among us, and that from our number the most deserving would be chosen as our generals. But what has actually been done?

In the name of God, Sir, does the history of the world present such another instance as that of our government, which, having at its disposal men trained to be soldiers from their boyhood, who were educated expressly for the Army in probably the best Military Academy in the world, passes over these men, leaving them in their time honored positions as Captains and Lieutenants, and goes behind the counter, in the county courthouses and low village barrooms to select her generals, her colonels, and other officers of her new regiments?

Such men as Duncan and Lee are kept down—they are sufficiently rewarded by an empty brevet, meaning nothing and doing them no good.

But Patterson, a rich old broker, Pillow, a twenty-fifth rate country lawyer, and such men are our generals!

But what else can be expected from such a President? I verily believe that Patterson and Pillow have humbugged the good people of the Union into the belief that they are the heroes of the War, that they and their precious bohunks and new levies have done all and everything!

I have served with both, and know that neither has done other than harm in the whole campaign. Patterson did nothing either at Vera Cruz or Cerro Gordo. Pillow, nothing at Vera Cruz and worse than nothing at Cero Gordo. I was so unfortunate as to be under his command at Cerro Gordo, and to his folly, his worse than puerile imbecility, the miscarriage on his side of the action is entirely to be attributed. This is the same man who attempts to wrest—to steal, I should say—from our Noble General Smith the credit of Contreras with which he had no more to do than you had.

Speaking of General Smith reminds me that it is a subject of wonder to the army how Polk could ever by any chance have hit upon such an appointment. It most assuredly must have been intended for someone else, for a better, a nobler man, never lived, and he is a soldier by nature.

Formerly the citizens' appointments had been confined to the Infantry and Dragoons, but of late they have been made in the Artillery, which is, or ought to be, a scientific Corps—those who were put in last year are below those of my class, but appointments have been made this year above the members of the graduating class. These gentlemen have not been in action, but certainly a man who has undergone the ordeal of West Point is entitled to rather more consideration in military matters than one of the same age who has not seen a military work in his life!

The same thing applies to the Infantry. Heaven alone knows how long my own Corps will remain untouched in this respect. The climax will be capped indeed when the Engineers are appointed from civilians.[4]

I think the affair will be complete when vacancies in the Engineer Corps are filled by civilians! With regard to this Rifle Regiment (Mounted), you are probably aware that when it was authorized there were applications from officers of the army, for appointment in it, more than enough to fill two such regiments. But one, a 2nd Lt., was confirmed as an officer of the Army, and with regard to him, President Polk remarked that if he were not a graduate of West Point, he should have had a Captaincy!

Nine Brevets were attached to this regiment; five men of my class—upon them you may easily imagine—fell the labor of drilling and disciplining of the companies. How they have acquitted themselves in these duties and in the field of battle, the conduct of the regiment demonstrates. Yet they—a part of them at least—are to be rewarded, and made to understand the worth of their services, by having a citizen placed over them, or it may be some Sergeant or Corporal who has done not one twentieth of what they have!

Ah, Sir, had it not been that the officers of our old army were made of the truest and sternest stuff, had they not been bound together by the firm tie of the Military Academy, you would have heard long before this of the light in which they view these things! As it is, they are writhing under the lash of contumely and insult heaped upon them by the Executive, though they know that every victory they gain will be claimed for others and made a pretext for more injustice, while they have fought and shall have conquered! The remedy for many of these things is in the hands of the Senate. I know, sir, that you, at least, will endeavour to apply it.

1. This is a draft of a letter to an unknown correspondent, though presumably intended for a political leader, perhaps Pennsylvania senator Daniel Sturgeon, who was a graduate of the Jefferson Medical College and therefore a former student of McClellan's father. The letter may or may not have been delivered but nevertheless expresses McClellan's contempt for volunteer officers and political appointments to army command.

2. Kit Carson, famed frontiersman and mountain man, guided the forces of Brig. Gen. Stephen Watts Kearney from New Mexico into California and was instrumental in extricating Kearney's command from the siege of Californio partisans under Don Andres Pico following the Battle of San Pasqual, fought near San Diego on 6 December 1846. For this service President Polk commissioned Carson a second lieutenant in the Mounted Riflemen, but Congress revoked his commission on 28 January 1848.

3. The Regiment of Mounted Riflemen, designed to be more mobile than infantry and endowed with greater firepower than cavalry, was created in December 1845 by the Twenty-ninth Congress. The Riflemen served under Col. Persifor F. Smith and distinguished themselves at Contreras and Chapultepec. Of the regiment, after the day's fighting at Contreras, Scott said: "Brave Rifles, veterans—you have been baptized in fire and blood and come out steel. Where bloody work was to be done, 'the Rifles' was the cry, and there they were."

4. This attitude was perfectly common among the regulars. McClellan's colleague Isaac I. Stevens, for example, wrote to his wife on 8 July 1847, "Our military establishment is so wretchedly organized that it is difficult for a man of acknowledged merit to rise." In the ten new regiments, "very few promotions were made from the existing organizations, in

consequence of which some of the ablest military men in our army see placed above them men totally devoid of capacity or zeal for the public service." Stevens, *Life of Isaac Stevens,* 158–59.

Diary Entry

CITY OF MEXICO, OPPOSITE ALAMEDA

3 NOVEMBER 1847

G. W. thinks that a captain will be sent out to command the Company, and that he (G. W.) will be relieved by the 1st March, 1848. Mc. thinks that no captain will come and that the unfortunate "duet" won't get out under a year, or longer. *Quien Sabe*?

Joseph G. Totten to Maj. John L. Smith

ENGINEER DEPARTMENT, WASHINGTON, D.C.

15 DECEMBER 1847[1]

Sir,

The Chief Engineer desires to express to you and request that you will cause to be announced to each of the officers of Engineering under your orders in the Army of Genl. Scott, mainly to

Capt. R. E. Lee
Capt. J. L. Mason
1st Lt. P. G. T. Beauregard
Lt. Isaac I. Stevens
Lt. Z. B. Tower
2nd Lt. G. W. Smith
. . . G. B. McClellan

. . . the feelings of pride and exultation with which he has noticed the great and brilliant services rendered by you and them respectively, in the successful, glorious achievements of the Army of the United States in its approach to and conquest of the Mexican Capital.

The concurring and willing testimony of all the commanders under whom the Engineer officers in that Army have served, places the conduct of these officers in positions so prominent as to entitle them not once only, but on several distinct occasions, to the special notice and reward of the Government.

Justly proud of belonging to a Corps which, distinguished for its zeal and ability in the conduct of peace, has added, on every opportunity of the present war, lustre to the Military character of the nation, the Chief Engineer for himself, and in behalf of all others of his brother Engineer officers, desires also to express to all the Engineer officers of the Army of Genl. Scott his gratitude for the lasting honor they have conferred upon their Corps, by their gallantry, zeal, and ability at the same time to offer, personally, his congratulations on the high personal distinction acquired by each with the consciousness of having rendered important services to their country.

The Chief Engineer has reason to know that the Engineer Company serving with the Army has, at all times, rendered important services, being distinguished for its courage and zeal, and exemplary in its discipline and subordination entirely supporting the talent and devotion of its officers and adding reputation and honor to the Corps.

J. G. Totten, Col.
Chief of Eng^rs

1. This letter from the Chief Engineer was published "for the information of the Engineer Troops" by order of Smith and McClellan on 12 January 1848. George B. McClellan Papers, Library of Congress.

George B. McClellan to John H. B. McClellan

MEXICO CITY
1 FEBRUARY 1848

My Dear John,

Everything is at a stand still among us here, except that once in a while some little expedition moves off, such as Gen. Lane's[1] to Orizaba after Santa Anna, and Col. Clarke's to Cuernavaca.[2] There are no predictions whatever of a movement upon Guanajuato and San Luis for some time to come. Indeed, it is not at all probable that we shall have force enough to make such a movement for some six months. Nothing can be done until Congress authorizes more troops, without the recruiting for the regular regiments now in service goes on much more rapidly than it has ever done before during the war.

It is, I suppose, needless to tell you how much our strength has been exaggerated ever since we left Vera Cruz. They are making a great deal of fuss about peace. Indeed, it seems to be a well authenticated fact that propositions have actually been agreed upon between Mr. Trist[3] and the Mexican commissioners and that these propositions have been sent on to Washington for ratification by our Senate. The Mexican Congress meets in three days at Queretaro and these same terms will be at once submitted to their consideration.

You will have seen before you receive this, the papers of Rosas, Pena y Pena's Secretary of State, in relation to the San Luis pronunciamento, in which he states that no propositions made by our government are at all dishonorable for Mexico. His whole manner breathes a desire to make peace. We have been deceived so very often in regard to peace and c. that I hate to think about it, much more to convince myself that peace will really be made. I can only say that I sincerely *hope* that they will make peace soon enough to let us get home before July or June. We will have all kinds of rumors and counter rumors for many a month to come. If we get home next winter, I shall think that we are very fortunate. I for one am tired of a war in which there is nothing to be gained except a broken head or a wooden leg.

I have been thinking lately about getting my little mare home. I wish to keep her with me at the North, for I have become too much attached to her to part with her if I can help it. As I have no means of ascertaining, please look on the map and tell me how far it is from New Orleans to Philadelphia, both via Nashville and direct. If I cannot possibly take her home, I shall leave her at Tom English's where I know she will be well taken care of. How much per month does it cost to keep a horse in Philadelphia and in the country about there? I am afraid that it would be almost too long a sea voyage for her from Vera Cruz to New York, although horses have gone from Baltimore to Corpus Christi in safety.

You know, I presume, that we have a couple of daily papers in full blast here. As one of them also publishes a weekly sheet, will send it on to you that you may form some idea of what we are about in this out of the way portion of the world. I sent by the last train (which ought to arrive at New Orleans today) some music for my future sister in law. Did you get it? I will be a perfect curiosity shop in a small way when I return, between Aztec antiques, blankets, spurs, bridles, bits, rag figures, and c.

February 22nd—Well, the 22nd is over (its 10 at night now), and a great time has there been. Smith's Brigade was reviewed today—fired blank cartridges—and then marched by our Old General's quarters. They made the city ring with their cheers when they marched by his house, as the fine old soldier came out on his balcony. The noble old fellow must have felt that even if the administration has relieved him from command, they could not weaken the hold that he has upon the respect and affection of every man in the army. I for one will never say another word against General Scott. You must imagine what utter disgust and anger we feel in having our old chief, under whom we have so often fought and conquered, taken away from us. I must not speak of it, for I cannot allow my mind to dwell upon the subject without becoming enraged. My God! The base idea of going into action under this Mohawk Butler[4] is awful. Heaven grant that not another shot may be fired!

There was a magnificent exhibition of fireworks from the Palace tonight. I don't believe that the Mexs ever saw the like before.

You know all about the treaty, I presume. I hope it may be confirmed on both sides, but I very much fear that it will not. There is so much *rascality* in both Congresses that I have no faith in them. I am thankful that I am a soldier and not a politician.

We will hardly be able (that is, the whole Army) to leave the country before the sickly season sets in. I think, though, that if peace be made, our company will get out in the spring. I hope so. We vegetate here in the same old way. I know more Mexican families now, so that I can kill time a little more easily. We still live in Lombardini's house.[5]

It is to be hoped that our sovereign Congress will, in the course of time, get through their important business of President making and remember that we are here and that six months have elapsed since battle. If they intend conferring brevets, it is fully time that they should be at it.

February 27th—I received Frederica's letter of the 23rd January on yesterday. I must acknowledge that it rather depressed my spirits. She tells me that you have made yourself liable for our poor Father's debts. I was in hopes that the horses, book, and c. would have proved sufficient to cancel all liabilities, but fate has decreed it otherwise and all that can now be done, my dear brother, is that we two, poor and unknown though we be, shall strain every nerve to clear off the estate.[6] Mother spoke in her last letter of a certain horse being kept for me. Do not keep him, but sell him and devote the proceeds to paying some of those debts. Much as I love my little mare who has more than once saved my life, I shall even sell her if I can get a very high price for her, though if I cannot do that, I will keep her.

I trust you know me better than to suppose that I will allow you to pay alone these debts. If our uncles have allowed our father to do so with regards to our grandfather, and to involve himself for them, I am determined that your children shall never cast such a reproach on me.[7]

I know that you are very busy, but I must beg you to take an hour from your sleep to write me the exact amount of the debts and how great a part that the book will cover. Tell me also how much Mother has a year at present and how much she will have when all her property is available. If you could know how anxious I am to know, how often and how much I am grieved about it, I am certain that you would no longer hesitate, but would at once let me know everything, for it will be a great relief to me to be no longer in this state of uncertainty and doubt.

Row Master Arthur up Salt River for not writing to me. Ask Mary if she can sing for me when I return.

1. Joseph Lane, colonel of the Second Indiana Infantry, was appointed brigadier general of volunteers on 1 July 1846 and assigned to the command of the Indiana Brigade. He was brevetted to major general on 9 October 1847 for his conduct at the Battle of Huamantla. Following his discharge, Lane received an appointment from President Polk as governor of the Oregon Territory, and when Oregon was admitted to the Union in 1859, he was elected as the state's first senator. In 1860 he ran as a Democratic candidate for vice president on John C. Breckinridge's ticket.

2. Newman S. Clarke, colonel of the Sixth Infantry, was brevetted to brigadier general for his "gallant and meritorious conduct" at the siege of Vera Cruz.

3. Nicholas P. Trist was a member of the West Point class of 1822 but did not graduate. Instead he studied law under Thomas Jefferson, whose granddaughter he married, and

during the Polk administration became chief clerk of the State Department. Polk sent Trist to arrange an armistice with Santa Anna in exchange for a payment of $10,000. He arrived at Vera Cruz on 6 May 1847 and opened negotiations, but when the Mexican president appropriated the money but failed to produce the truce, Polk, disgusted by his envoy's seeming incompetence, ordered his return to the United States. Trist, however, ignored the president's instructions and successfully negotiated the Treaty of Guadalupe Hidalgo. For defying the president, Trist was dismissed from his post. He died on 11 February 1874 in obscurity and relative poverty. Wallace Ohrt, *Defiant Peacemaker: Nicholas Trist in the Mexican War* (College Station: Texas A&M University Press, 1997).

4. William Orlando Butler, a native of Kentucky, rose from the ranks in the War of 1812 to received a major general's brevet for his conduct at the Battle of New Orleans. Returning to civilian life, he practiced law and served two terms in the House of Representatives as a Democrat. With the coming of the war with Mexico, he was commissioned a major general and commanded one of Zachary Taylor's three divisions at the Battle of Monterey, where he was wounded. After recovery, he returned to Mexico late in 1847 and assumed command of the army at Mexico City when Scott was recalled to face a court of inquiry.

5. Gen. of Brigade Manuel María Lombardini was severely wounded at Buena Vista but by September 1847 was sufficiently recovered to take part in the defense of Mexico City. Upon the fall of the capital and Santa Anna's subsequent abdication, Lombardini took command of the army. For a brief period in 1853, he was head of the national government. His home was located at No. 2 Calle San Francisco.

6. At the time of his death, Dr. George McClellan was working on a medical textbook that was finished by his son Dr. John Hill Brinton McClellan and published as *Principles and Practice of Surgery* (Philadelphia, 1848).

7. One of McClellan's uncles, Joseph, was the father of Henry Brainerd McClellan, future chief of staff to Confederate generals James E. B. Stuart and Wade Hampton and author of *Life and Campaigns of Major-General J. E. B. Stuart, Commander of the Cavalry of the Army of Northern Virginia.*

Extract from Engineer Orders No. 1

WASHINGTON, D.C.

1 MARCH 1848

II. Cap[t] Geo. B. McClellan, Corps of Engineers, is assigned to duty at the Military Academy, West Point, as Instructor of Practical Engineering, and as Treasurer, and will take charge of the construction of the Cadet Barracks; forthwith relieving Cap[t] F. A. Smith, from all his public trusts and duties at that place.

George B. McClellan to Elizabeth Brinton McClellan
CITY OF MEXICO
22 MARCH 1848

I suppose, my dear Mother, that you can scarcely imagine us roasting at this season of the year, but nevertheless, it is very hot, so much so that I should be quite glad to think that we could get out of this country before the sun gets North of us again. We are expecting the treaty from Washington, either ratified or rejected, every day, allowing the Senate eight days to make up their minds, seven days to New Orleans, and seven more to Vera Cruz, it should have arrived at the latter place on the day before yesterday, and be here about Saturday.

There appears to be no doubt that a quorum of the Mexican Congress will convene as soon as we hear of the treaty. It would be useless for them to meet before, as they cannot act until the action at Washington is known. Should everything progress most favorably (but I cannot believe that it will be so) we may leave here in three weeks. Our company will leave with the first who start from this city, but I shall consider myself very fortunate if we get out of the country before next winter. I really do not expect it, for it is but fair to suppose that many unlooked for delays will arise to detain us. If we do get off in a month from now, I will get to Philadelphia early in June, but I repeat it that I by no means expect to leave this city before next winter.

I am almost ashamed to confess it, but now that there appears to be a probability of going home, I cannot help feeling a little sorry to leave this city. I most certainly do feel, and shall be ashamed not to, very sorry to think of the associations that must be severed when this army is broken up and scattered over the length and breadth of the U.S. The attachments formed in a campaign with men by whose side you have fought, who have shared with you hunger, thirst, fatigue and privations, with whom you have exalted in the moment of victory, and when you have met day after day such attachments cannot be severed without a pang.

The Court of Inquiry is now [in] session—the accusations of that ungrateful Worth against General Scott are villainous, also those of Scott against Col. Duncan, but the Pillow case is now under investigation. Our noble old General Scott is coming out of it most magnificently. Pillow is getting mired worse and worse every day. He has already been proved to be a liar, a scoundrel and a mean, pitiful rogue. You cannot imagine the feeling in the old regular army in reference to General Scott. Today, before his conduct with reference to this persecution of the administration, I never thought him to be a great man, but now I do believe that he is really a very great man. I will preserve the proceedings as published in the daily papers here so that when I return you may see them in full. The old gentleman will floor them all.[1]

In the letter (scrawl, rather) which McAllister[2] took for me the other day, I forgot to say that I returned last week from an excursion to the famous mines of Real del Monte, about 70 miles from here. They are silver mines. One shaft is carried to the depth of a little over 1500 feet. We descended a shaft not so deep, but the richest one now worked. They are worked by an English company. The Englishmen treated us with the greatest hospitality, gave us fine old English dinners and c. We went also to the Hacienda of *Regla,* about fourteen miles from the mines. At this place there is a formation of basaltic columns superior in height, if I remember correctly, to those of the Giant's Causeway Fengal's Cave.[3] They are 210 feet high. A beautiful stream forms a fine cascade in the midst of them.

Then too we took dinner with the gentlemen. There is here an English lady who seemed quite well pleased to see so many white (we were rather red I believe I must confess) faces around her, and to hear the sound of the English tongue.

We think of getting up an expedition to visit the top of Popocatepetl. If we do not go too soon, we will most certainly do so. Few persons have ever been there.

On the day after tomorrow it will be a year and a half since we left West Point. How the time has flown! I have seen a great deal in that year and a half—more, probably, than I will see in all the rest of my life. Well, I suppose I must make up my mind to pass the rest of my life in a very tame and humdrum manner. I have no doubt that I will often wish for the excitement of a campaign—heigh ho!

It appears intended that G. W. is to retain command of the company when we return home. In that case, I shall probably remain with it for a long time. I should prefer it greatly to being in a bank. I should like the duty and like the States, for, after all, West Point is a fine old place, much as I did hate it when I was a Cadet.

If I do get home in June, you and the children must go on there and stay some time with me. We have been ordered to move out of this house. The commissioners asked as a particular favor that General Butler would have this house vacated, it being that of Gen. Lombardini, Commander in Chief of the Mexican Army.

I don't know where we will go to now. We won't have such fine quarters as these, I am quite certain. I hope, however, that our stay here will be so short that it will make no difference as to where we live for the nonce. If peace is not made, we may get the house back again. The worst of it is that Lombardini is a great rascal. He maltreated our prisoners very much, so that we owe him no kindness whatsoever. We have the consolation of knowing that we have enjoyed ourselves whilst here, and have broken some of his chairs for him, anyhow.

Well, my dear Mother, I hope soon to see you again and present to your admiring view the sunburned face of your hopeless scapegrace of a son. Kiss Arthur and Mary for me. Tell John that I have given up all hopes of every hearing from him. He is incorrigible, the fine fellow. Give my love to Frederica and the children. Remember me to "all concerned." Do write soon to your affectionate son,
Geo. B. McC

G. W. sends his kindest regards. Have you written to Madam "G. W."?

1. Fearful that Scott's victories in Mexico would win for the general the presidential election of 1848, Polk recalled Scott, an ardent Whig, from his command and subjected him to a court of inquiry on highly questionable charges of unlawfully arresting the manifestly incompetent and contemptible Gideon Pillow and of conspiring to bribe Santa Anna to end the war. Scott was cleared of all charges. Bauer, *Mexican War,* 370–74.

2. Julian McAllister, a member of the class of 1847, served through the Mexican War in the Second Artillery but in 1848 was transferred to the Ordnance Corps. He died in 1887 as senior colonel of the branch. *New York Times,* 4 Jan. 1887.

3. Fingal's Cave, located on the island of Staffa off the west coast of Scotland, was discovered by Sir Joseph Banks in August 1772. It was the inspiration for Felix Mendelssohn's *Hebridean Overture* and was the subject of an 1832 painting by J. M. W. Turner. Sir Walter Scott described this famed sea cave as "one of the most extraordinary places I ever beheld."

Diary Entry

POST OFFICE, MEXICO CITY

15 APRIL 1848

Captain hasn't "arrivo"—duet still here—year most half out and ain't off yet!!!

George B. McClellan to Arthur McClellan

CITY OF MEXICO

20 MAY 1848

My dear Arthur,

I was very much pleased at receiving your letter at the last mail. I was very much surprised that you could do so well—go on and keep it up. As to that *rat* that appears to trouble you so much, why, we must try to give him a "right against Infantry, in tierce, cut." If that don't settle him, we'll chase him with my lance. Between the two, I'm inclined to believe we'll get him at last.

I am glad to hear that you are going to school and improving so much. Tell Mother that the best recipe I know of for making Mary fat is that you should all spend the summer and fall with me at West Point. I think a turn in the mountain air would benefit you all greatly. What do you say to it, "muchacho"?

Tell Mother that a little bird whispered in your ear that I shall be home in two months or less, and that if I do not remain on duty with this Company, I shall be stationed at the Pea Patch. I have your rag figures all safe and shall bring them with me shortly. I also have some wax figures to bring you.

I have a message for you and Mary from one of the handsomest ladies in the city—I'll give it to you when I return.

You must be looking forward to the summer holidays soon—when do they begin?

Have you made up your mind yet whether you will be a Bishop, a Mechanist, a Lawyer, or a Soldier? Don't be in a hurry about it, but take plenty of time to make up your mind. In regard to the last Profession,

I can tell you from some little experience that a soldier gets more hard knocks than sixpences—so dodge that, anyhow.

I wish you were here to ride with me on the Paseo de la Vega this afternoon. A beautiful canal runs along its side, the country around consists of finely cultivated gardens, and the Paseo itself is filled with carriages and horsemen. It is a very handsome sight.

Now, young man, as I've nothing more to say, and a great deal to do and very little time to do it in, I will take the liberty of saying Adios to you—thanking you for your condescension in writing to me, and I have the honor to remain
your affectionate brother
Geo. B. McClellan
Lt. of Engrs.

Kiss Mary for me—give my love to Mother and James.
Hasta luego.
I've taken the liberty of enclosing in the same envelope a note for Mother—please excuse me.
Mr. Smith sends his respects to Mr. Arthur and hopes soon to see him.

George B. McClellan to Thomas C. English

MEXICO CITY
23 MAY 1848

My dear Tom,

We are busy now in making our preparations to start for the States. General Smith goes to Vera Cruz tomorrow in order to superintend the preparations in the way of transportation &c.

When Songo came out, he handed me a memorandum of yours saying that $159 wages were due him. He asked me this morning to write to you, requesting you to send it, or such part of it as may be convenient, to my order. It had better be sent to me at Philadelphia, to the care of John, and it will remain there until I arrive. I expect to get there about the beginning of July.

The lower House has ratified the treaty and there is not the slightest doubt but that it will be ratified by the Senate. The Mexicans will have grand times when we have [gone]—Theatre, balls and Pronunciamentos. Well! if they will only wait until we have embarked, they may pronounce as soon as they like. I suppose Uncle Sam will think twice before he meddles in Mexican affairs again.

You must excuse the brevity of this, but I am up to my neck in work, receiving the Engineer and Company property, for Smith G. W. goes with General Smith tomorrow, leaving the command of the Company with me. In haste, I remain

Truly yours

Geo. B. McClellan

Lt. of Engrs

Robert E. Lee to George B. McClellan

VERA CRUZ

6 JUNE 1848

Sir,

The Services of the Engineer Company being no longer required with the army in Mexico, and in compliance with instructions rec^{d} from Major Genl. Butler, you will repair with it to West Point, New York, with such of the company property as may be in your possession and report its return by letter to the Chief Engineer at Washington and your arrival at West Point to the Superintendent of the Military Academy.

Genl. Smith, commanding this Department, has already given orders for the embarkation of the Company on board the Brig *Helen* bound for New Orleans. On your arrival in that city, you will apply to the Quarter Master of the Post for transportation to West Point.

Very respectfully, your servant

R. E. Lee

Capt. and Sen. Engr. in the field

7

"THE ENGINEER COMPANY IS DESTROYED"

WEST POINT

23 June 1848–22 September 1849

The end of active campaigning very nearly brought an end to McClellan's military career. The young soldier, who had only months before written to his mother from Mexico City, "*Aint* I glad I'm alive? I tell you that one can't tell what fine fun it is to live until he has been through about six battles," was now nearly despondent. His letters from West Point in the months following his return reveal a junior officer increasingly disillusioned with the monotony of garrison life, the political neglect or unwonted interference with the military establishment, and the slow rate of promotion in the peacetime army. Moreover the pomp and circumstance of glorious war that had so filled the head of Cadet McClellan was beginning to lose its glitter. When a committee of the first citizens of Philadelphia subscripted to a presentation sword for the newly returned lieutenant, McClellan wrote to his mother: "They will humbug me into the belief that I am somebody, whereas I well know that [it] is best for one not to get too proud for his place. Sincerely, I would rather they would let me alone when I don't have a chance of doing anything that deserves notice." While admitting that he would be pleased to receive the sword, he also confessed, "it actually causes a kind of feeling of shame to arise in me, thus being rewarded for doing a subaltern's duty in such a small business as the Mexican War!" Only an appointment as engineering officer on the Red River exploring expedition of 1852, led by his future father-in-law, Randolph B. Marcy, induced him to remain in the army.[1]

But the war with Mexico had done much to ameliorate the hostility that many Americans of the Jacksonian persuasion had felt toward West Point–trained officers and toward the regular army in general. As for the sappers and miners, Chief Engineer Joseph Totten stated that the opportunities of service in the company of engineers "have been profited of, by the sergeants and rank and file, as well as by the commissioned officers, to display the highest qualities as soldiers, demonstrating, at the same time, the great advantage to armies, however engaged in the field, of possessing troops well grounded in the peculiar exercises of engineer soldiers."[2]

From among the noncommissioned officers of the company, Thornley S. Everett, the company's property sergeant, was brevetted a second lieutenant in the First Artillery, 28 June 1848, and rose to the rank of captain before dying as a prisoner of war at Andersonville, Georgia, in April 1864. First Sgt. David H. Hastings, who G. W. Smith described as "a well educated man, very intelligent," and "a remarkably fine looking soldier," was brevetted a second lieutenant in the Second Dragoons, 28 June 1848, and rose to the rank of major of the Fifth Cavalry before he retired, 7 December 1863. And Sgt. Samuel Henry Starr was brevetted a second lieutenant in the Second Dragoons, 28 June 1848, becoming colonel of the Fifth New Jersey Infantry at the outbreak of the Civil War. On 25 April 1863 he resigned from the volunteers to become major of the Sixth U.S. Cavalry, from which he retired with the rank of colonel. In addition two other sergeants, five corporals, and twenty-two privates of the company were recommended for certificates of merit.[3]

As early as 1847 Totten had urged the president to create "two or three additional companies, to be officered, as the present company is, from the corps of engineers," and at the war's end, he again recommended that the company of engineer soldiers be expanded to battalion strength. Secretary of War [William L.] Marcy, generally no great friend of the professional soldier, reported to Polk that the company had "rendered efficient service" and concurred with the chief engineer's proposal for "an increase of this description of force, as required to meet the wants of our armies in the field." Nevertheless within six months the veteran engineer soldiers were discharged by an act of Congress, reducing Company A to a squad of recruits and leaving McClellan to seek active service on the western frontier.[4]

1. George B. McClellan to Elizabeth Brinton McClellan, West Point, 19 Jan. 1849, George B. McClellan Papers, Library of Congress.

2. *Annual Report of the Chief Engineer to the Secretary of War,* 30th Cong., 1st sess., 1848, S. Exec. Doc. 4, 628.

3. Joseph G. W. Smith to Joseph Totten, 4 May 1847; Joseph G. Totten to William L. Marcy, "Report of the Chief Engineer," in *Report of the Secretary of War,* 31st Cong., 1st sess., 1849, H. Exec. Doc. 1, 281.

4. Totten to Marcy, "Report of the Chief Engineer," *Report of the Secretary of War,* 29th Cong., 1st sess., 1846, S. Exec. Doc. 1, 137; *Report of the Secretary of War,* 30th Cong., 1st sess., 1847, H. Exec. Doc. 8, 67; Totten to Marcy, "Report of the Chief Engineer," in *Report of the Secretary of War,* 31st Cong., 1st sess., 1849, H. Exec. Doc. 1, 281; Smith, *Company "A,"* 35–36, 44, 63, 72, 79–81, 90.

George B. McClellan to Joseph G. Totten

WEST POINT

23 JUNE 1848

Sir:

I left the city of Mexico on the 28th of May and reached the Post at 6 P.M. of yesterday, the 22nd. I find no preparations made for the accommodation of the men. The Barracks occupied by the detachment of recruits are altogether insufficient for the whole command; the men of my company are now in the porch of the Barracks.

I expected to find some orders or directions awaiting me as to the disposition of my company and myself. On the contrary, both Captain Cullum and myself found ourselves in a very disagreeable uncertainty as to our relative positions, which has been put an end to by the following order of Captain Brewerton, the Superintendent.[1]

Premising that I neither expected nor desired to retain command of the Company for a period longer than that necessary to receive instructions from the Department on the subject and that I confidently anticipated finding orders at West Point for me to turn over the command of the Company to Captain Cullum or some other *Captain* of Engineers—I most respectfully protest against this last order as being issued without the proper authority, because I was ordered to report merely the *fact of my arrival* to the Superintendent and to the Chief Engineer for orders. As I understand the matter, the Superintendent could issue no such order on the subject except after receiving instructions from the Engineering Department, for I do not consider myself as having reported to him *for duty.*

I would most respectfully request instructions on the subject at the earliest convenience of the Department.

By authority of Captain Lee, I sold the greatest part of the Engineering Train belonging to the Company in the City of Mexico for $770—being the then list prices as nearly as we could ascertain them. I have also to request [instructions] with respect to the disposition of this money which is now in my hands. The proper papers in the case will be forwarded immediately upon the arrival at this post of Sergeant Everett who has in his possession the receipts, and Sergeant Everett obtained a furlough at New Orleans until the last of the month.

The Company arrived in good condition. No man died on the passage.
Very Respectfully &c.
[George B. McClellan]

1. Henry Brewerton, an 1813 graduate of West Point, served with the Corps of Engineers through 7 March 1867, achieving the rank of colonel and brevet brigadier general.

Head Quarters, Military Academy

ADJUTANT OFFICE, WEST POINT, N.Y.

23 JUNE 1848

SPECIAL ORDERS

NO. 92

2nd Lieut. G. B. McClellan, Corps of Engineers, having returned from active service in Mexico with the remnant of the Engineer Company, and reported to the Superintendent in conformity with orders dated the 6th inst., received from Captain R. Lee, Chief Engineer in the field, will turn

over the command and all company paper and property to Capt. G. W. Cullum, Corps of Engineers, to whom he will report for duty.
By order of Captain Brewerton
J. M. Jones[1]
1st Lt., 7th Inf.
Act. Adj., M.A.

1. John Marshall Jones, class of 1841, was assigned to the academy as an instructor and so did not participate in the Mexican War. Commanding a brigade in the Army of Northern Virginia, he was killed in action during the Battle of the Wilderness, 5 May 1864.

George B. McClellan to Frederica M. English

WEST POINT
25 SEPTEMBER 1848

My Dear Sister,

I fear you think either that I am most terribly lazy (which is pretty near the truth) or that I have forgotten you (which is pretty far from the truth)—the fact is that ever since I returned from Mexico, I have been knocking around, first to Philadelphia, then to New York, then to Philadelphia, then to Boston, then to Washington, &c., &c., so that I have had no time for anything.

I got back from Philadelphia two weeks ago and as the Hotel [Cozzens's] has been crowded until this morning, I have been occupied in doing up the polite. This morning, however, the premises were abandoned and we are left pretty much alone in our glory, so I shall try now to sober down to a quiet course of life.

It may be a matter of some interest to you to know that I am doing my best to go to Oregon or California. I am in [illegible] infected with the "furor viande." I cannot rest contented until I have seen some little of this big world of ours. I have just seen enough to cause me to wish to see more, and being so fortunate as to be neither engaged nor in love, I have nothing to keep me back from indulging the bent of my whim to the fullest possible extent.

I have applied to go to Oregon with General [Persifor] Smith and shall go if any Engineer Officer goes with him. I fear, tho', that none will be sent, for a couple of years yet. Failing in that, I shall try to go out with the board of Engineers. If I fail in both, I shall be very much disgusted.

Will you believe it? I am really and sincerely tired of this part of the world. I long to be in the woods again and on the march. You cannot imagine the pleasure and contentment of a march. You have constant change of scenes, of faces, of fare, and everything. I really believe that I was intended by nature to be a rich man. If I were, I'd start on my travels at once. I am afraid that I am doomed to disappointment about the Oregon and California business. It will be just my luck (as everyone says who don't get everything he wants).

*****

You will be astonished to hear that Songo, the faithful Songo, has so far forgotten his dignity as to get married. Yes, it is a melancholy truth, that Songo was married when we were last in Philadelphia. Not only that, but he married an oyster seller besides—lucky dog! His master sincerely wishes that he may be so lucky (if ever so unfortunate as to be married at all) as to marry as much. Just think of it!—Songo behind a counter, decked out in all the splendor of a clean shirt and white apron, flourishing the oyster knife, bestowing benignant smiles on all new comers, chuckling to himself as he drops in the drawer the ever flowing sixpences, praising his oysters, thanking Heaven that he is no longer the servant of an officer in Mexico, and (I hope) sending McClellan oysters at half price with the crackers thrown in (not the plate, but *gratis*)!

Songo is still with me here, but I shall let him go to his dear Andromache[1] (or Dinah) pretty soon. I am very sorry to part with him, for he has always been an excellent, honest servant to me. He tells me that he would like to have the $109.00 that Tom still owes him, as soon as convenient. He wishes to invest it in the oyster business with "father in law." When you send it, enclose it to me. I believe that I told you that I had received the $50 you sent when I was in Philadelphia.

There is nothing new in this part of the world. Everything goes on in the old channel. Even the Presidential election does not seem to create much of an excitement. I hope Old Zack gets it.

The Point looks beautifully just now. We have had so much rain during the last two weeks that everything looks fresh again. Jack Frost will soon be on us again. It will be quite a novelty to me to feel a good cold winter. Love to Tom and the children. Do write often to me and believe me to be
your affectionate brother
Geo B. McC

1. Andromache was the wife of the Trojan hero Hector.

George B. McClellan to Arthur and Mary McClellan

WEST POINT
24 DECEMBER 1848

My Dear Children,

A Merry Christmas to you and plenty of them is the sincere wish of your brother who regrets very much that he cannot spend this Christmas with you. What do you think of a Profession that has kept me from home for seven Christmases in succession, and which causes one to move so much from one place to another that I have no idea whether I shall be at West Point, Philadelphia, California, or where the next Christmas? I hope that I shall be with you next year, but cannot tell.

It is a pleasant life tho', to be constantly changing and to see so many strange places of which one reads. Would not you like to see the countries and cities you read of in your geographies?

Three years ago I was a Cadet, and hence, two years ago I was marching over the hot prairies, as hot as it is in Philadelphia in August, nothing to drink but muddy water. Last year I was in a fine house belonging to a Mexican General, in the city of Mexico, with every comfort, and a delightful, warm climate, like our spring weather, no snow on the ground and no ice, and I could not tell then where I should be now. This year I am here in the midst of snow and ice, the snow is almost a foot deep on the ground and we have elegant sleighing, but it is very different from Mexico. My nose and ears were most frozen off in walking down to inspect the Company just now.

The Cadets had what they thought a fine supper last night. Poor fellows, they don't get many of them and they enjoyed it very much, I imagine—at least I remember that I used to think them very fine.[1]

Talking of Cadets, tell John to say to Dr. Morton that I received last night the package for his son and gave it to him this morning.[2]

*****

Again wishing a Merry Christmas to you,
believe me to be
your affectionate brother,
Geo B. McC

1. The regulation bill of fare at the academy was an unvarying round of boiled meats, starches, boiled vegetables, cornbread, molasses, and with breakfast, "radishes in the season of them." But according to the 1848 board of visitors' report, "The corps are furnished a poultry dinner on thanksgiving, Christmas, and new year's day; also, on the eve of the two last mentioned days, an oyster supper." As John C. Tidball wrote, at Christmas "came a dinner of good, fat turkey, a most grateful break in the monotony of our plain mess hall fare. . . . Most gorgingly did we enjoy it." *Report of the Secretary of War,* 31st Cong., 1st sess., 1849, H. Exec. Doc. 1, 300; James L. Morrison Jr., ed., "Getting through West Point: The Cadet Memoirs of John C. Tidball, Class of 1848," *Civil War History* 26, no. 4 (Dec. 1980): 318.

2. James St. Clair Morton of Philadelphia was graduated second in the West Point class of 1851 and was commissioned into the Corps of Engineers.

George B. McClellan to Maria Eldredge McClellan

WEST POINT

JANUARY 1849

My dear Maria,

Three days ago I commenced a letter to you which I was obliged to leave unfinished so I tore it up, and here goes another one, which I suppose will meet the same fate, but I won't destroy it. You would be perfectly justifiable in never writing to me, had I no excuse for my long delay in answering letters, but I can't help it.

I was just now talking to an officer who leaves here on Monday en route for Europe for the benefit of his health, and the idea of not being able to go myself threw me into such a fit of disgust that I have hardly recovered from it yet—but if I ever do get out of debt and a little ahead of my pay, the first thing that I shall do will be to get a leave of absence for a year and start for the other side of the water. I would like to go now, in order to see the Austrian Army in the field in Hungary,[1] but as it is quite probable that they will amuse themselves by fighting for some years in some part of Europe, I suppose I shall see something of the kind whenever I do go—if I ever do.

So Louis Napoleon is President! Did you see the account of his grand review? It must have been a splendid sight. What recollections it must have aroused in the old Invalids—to see a Bonaparte again at the head of France.

Well, I got so far this afternoon, and as I have to sit up until after 12 in order to go down to Barracks and visit the guard, being officer of the day, I don't know what better thing I can do than try to finish this.

*****

We are still hammering on here in the same dull routine of duty—no change, no variety. The river is again open and our skating spoiled, so I anxiously wait for spring when recitations will cease and we can get at drills again—anything for a change. You can't imagine how dull the monotony of a garrison life is after a year or two in campaign—yet I have plenty to do and am generally quite contented tho' I occasionally get a "big disgust" and growl for a while until I get pleased again.

The last two or three days I have been in a grumbling fit, perhaps I'll get over it by tomorrow. The walk down to Barracks in the cold may restore my equanimity. But I never expect to have such fine times again as we used to have on the march in Mexico—constant change, constant excitement—if we had no battle, we at least had bull fights—if we could not go to a Protestant Church on Sunday morning, we could go to the theatre on Sunday night—so it went, and alas, it is gone sure enough, probably never to return. The recollection and the "caravansaries" are all alone left. My gracious, what pleasure it is to get with some comrade of the war and talk ("gas," as we say) over old times—the lordship we had—the narrow escapes—the beautiful views, &c., ad infinitum.

I'm getting out of my disgust already. I think that if I [were] ever at home I could soon talk myself into a good humor, provided you could master patience enough to listen to me. But here I am, alone, in an old North Barracks room, a hideous Mexican God on one side, an equally beautiful picture on the other, a suit of armor, with a pair of skates hanging over it on one shelf—a lance in a corner—my old sabre hanging up by way of setting off sundry rapiers, foils, masks, shell jackets, cits' hats, &c. and now you may form some idea of my room.

*****

Your affectionate brother
Geo B. Mc

1. In 1847 Hungary attempted to free itself from the domination of the Austrian Empire, but after two years of bitter fighting, the rebel army was routed at Temesvar, 9 August 1849, in the last great battle of the unsuccessful war.

Not until April 1855 did the U.S. Army send McClellan to Europe for a year as a military observer, detached to observe and report on the Crimean War.

George B. McClellan to John H. B. McClellan

WEST POINT
10 JANUARY 1849

My Dear John,

I don't know whether it is the fault of the mails or that you all have concluded to punish me for not coming home to spend Christmas with you by not writing to me. It was very fortunate that I did not go home, for I should have been ordered back as soon as I arrived.

On Christmas, orders were received here from the Chief Engineer requiring plans and estimates for several buildings to be furnished him for the Military Committee of the House by today at the latest. Among those required was a Barracks for our Company, and I had to make all the drawings. The Barracks had to be planned and drawn in the short time allotted, and from two weeks from today until last Saturday night at 12 o'clock I drew every day, morning, afternoon, and night, working

Sundays, New Year's day, and all. I had to make eight different drawings on the same large sheet, fifty-two inches by thirty-four inches, all drawn accurately to a scale, all the details, and painted, so you may imagine that I had my hands full. I hope that Congress will pass the appropriations, for we cannot get on another winter without the new Barracks.[1] Our men are now anything but comfortable.

*****

We had quite a performance here on New Year's Day. All the colors taken in the Mexican War were sent here to be kept, and they were received with the due military honors on that day. The Corps of Cadets and our Company marched down through the snow to receive them. We carried the two colors taken by us. You perhaps saw an account of it in Monday's *Herald*.[2]

I received and delivered Morton's package. I saw him not ten minutes ago. He has just got through his examination in Mathematics and will, I believe, be 2nd, as before. He is a very fine fellow, indeed. He is 1st in drawing.

What kind of weather have you had? It has been very cold here and we have had fine sleighing for about three weeks.

I received a letter from G. W. last week in which he and the Madam asked to be remembered to all of you, including Maria. He asked me to congratulate you. They had arrived safely in Mississippi, all well.

I do not think that I shall make any further application for California. They may do what they please about it at Washington, I have said my say.

Tell Arthur that I dreamed last night that I saw him thrash a couple of youngsters for saying that he had a dirty jacket on. How is the youngster?

I had another funny dream the other night. I thought that I was home and that we gave a dinner party to some twenty persons and that upon going into the dining room we found nothing but my old mess chest with its tin plates and cups, with real camp dinner of pork and beans and hard bread.

Tell Mother I'll answer her letter tomorrow, but that she must not wait for answers for me, that my time is so much eat up, that I have hardly any leisure. Tell Maria that she must write, too.

I laughed a great deal when I received Morton's package, at the idea of having requested a demijohn of whiskey to be sent in it, but I thought it would be a box—*n'importante.*

The Irish laborers on the rail road opposite here have been playing the very devil. They have murdered an overseer or two and killed some dozen or so of each other. We were in hopes that the Company would be sent for to thrash them. I anticipated some very pretty shooting, but it appears that the riot has been put down.

So Louis Napoleon has been elected President of France. So much for a military name.

I wrote to Maria and to Arthur about Christmas day. Were the letters received?

I received last week a letter from Mr. Gardette asking me in what battles I was present.[3] What was it for? I answered it at once, of course.

Give my best love to Mother, Maria, Arthur, and Mary and believe me, my dear John, to be your affectionate brother,
Geo B. McC

1. The board of visitors that examined the academy that year was in entire agreement. "The barracks for the engineer company," they reported, "are quite insufficient. There are but three rooms, twenty-five feet square, now available for the accommodation of the 100 men composing the company of the sappers, miners, and pontoniers; and the arrangements for messing are equally insufficient. The position of these quarters is also extremely objectionable, being within one hundred yards of the post magazine containing a large quantity of powder." Colonel Totten concurred with the board's assessment, reporting to the secretary of war that "[t]he return of the company of engineer soldiers from Mexico, where their services were alike distinguished and valuable, and their increasing ranks by enlistment, make provision for their barrack accommodation indispensable," requesting $5,000 for the purpose. *Report of the Secretary of War,* 31st Cong., 1st sess., 1849, H. Exec. Doc. 1, 274, 291.

2. A correspondent for the *New York Herald* reported that the battalion of cadets, the company of sappers and miners, and Bvt. Maj. William H. Shover's company of light artillery marched down to the wharf, where they received thirty-two captured flags. In the words of the reporter, the flags, "bearing the marks of war's rough usage, torn by shot, discolored by sun and storm, and recalling most vivid associations of blood and battle, were thickly sown among the files of cadets," but the engineers, "whose name is closely connected with grape and canister, and the hand to hand struggle in the van of battle, bore in triumph the banners which they captured in Mexico." One of those flags, that of the Twelfth Regiment of Artillery had been captured at the Battle of Contreras by Artificer N. S. Read.

New York Herald, 8 Jan. 1849; G. W. Smith to W. T. H. Brooks, 23 Aug. 1847, in *Report of the Secretary of War,* 30th Cong., 1st sess., 1847, H. Exec. Doc. 8, 69.

3. Emile B. Gardette was a Philadelphia physician and for a time president of the Jefferson Medical College Board of Trustees.

George B. McClellan to Joseph G. Totten

WEST POINT
17 FEBRUARY 1849

Dear Sir,

It has just become known to me that some, or all, of the men of the Engineer Company with which I am still serving, who served during the Mexican War, have been making great efforts to obtain a general discharge, similar to that granted to the Rifle Regiment. That the men of this company have at least an equal claim with the Rifles to the favor, as far as concerns toilsome and gallant service in the field, I shall ever be the last to deny, but I understand that they have endeavored to procure their discharge on other grounds, that are by no means tenable.

First—They say that promises in relation to their duties were made which were never fulfilled, such as that they should not carry muskets nor march on foot, etc., etc. Whence these notions came, I know not; I know that the late Captain Swift, the first commander of the Company, was ever careful to put the worst light on the duties of the men before he would enlist them, for he was one of those rare men in whose nature deception did not exist. The present Commanding Officer, Captain Cullum, was careful to follow the same course, and by these two officers the greater part of the old company were enlisted.

I know that there were advertisements published in reference to enlisting for the Company, by another officer which were rather in the bombastic vein, and which might have misled men of ardent imaginations, but that that officer would ever enlist a man, knowing that he entertained false comprehensions, my knowledge of his high and honorable character forbids me for a moment to believe. The nature of his advertisements was, to the best of my recollection, about this: He called attention to the fact that the duties of Engineer Troops were ever of the most dangerous class and that consequently more numerous opportunities of distinction and promotion would be afforded in this company than in any other, especially as it was intended that it shall be composed entirely of Americans and those, too, of a respectable class. He mentioned, too, some instances of distinguished Generals who had risen from the ranks.

His advertisements might be charged with being faulty as to style, but no more. In reference to these hopes of promotion held out, they were actually fulfilled, for three Sergeants of the Company were made Brevet 2nd Lieutenants in the old army upon the recommendation of the officers of the company and four others, members of the company, were appointed in the ten new Regiments without such recommendations. Can the same be said of any other company?

The pay of these men is much better than that of any other troops, because of the higher order of intelligence and more dangerous and toilsome service exacted of them. They are fed and clothed equally well with other troops. They are subject at present to some inconvenience as far as regards quarters, but this is only a temporary evil pending the erection of proper barracks for the company.

With reference to the education promised them—as a matter of course nothing of that kind could be expected in the field, yet while we were in the City of Mexico, besides theoretical instruction received by them in the proper duties as Engineer Troops, they were taught arithmetic, also to read and write, and during the whole of this winter they have received instruction daily from their officers, so that I do not hesitate to say that the men who enlist or have enlisted in this company will, at the expiration of enlistments, have received an education that men of their class can receive at no other place in the U.S., for at what other place can men of this kind have graduates of the Military Academy for their instructors?

With reference to the period of their enlistment, they were enlisted, *bonâ fidê,* for five years and no hope or hint was given of a discharge at an earlier period.

If other reasons than these I hear maintained are given as grounds for their discharge, I know not of them, and I beg you to take it for granted that I fail to touch upon because I am not aware of their existence, not that I fear to handle them for I doubt not that they can all be proved "air, thin air."[1]

I understand that the men have great hopes of this bill being tacked onto the appropriation bill in the Senate and that Mr. Calhoun, Mr. Cass, and Mr. Dix have promised to vote for it.[2] Now, although I am aware that you have more important affairs with which to occupy yourself than the concerns of a company of troops, I thought it my duty to put the truth in your possession that you might make such use of it as you think fit. I may also state that if this bill passes, the efficiency of the company will receive a blow from which it can not recover for many years, because all our noncommissioned officers, with two worthless exceptions, are of the old war men, and we have none among the recruits at all fitted to supply their places. Ought, then, the good of the service to be sacrificed to the whim of a few of [the] men who have the California fever?

1. McClellan is here quoting Prospero from Shakespeare's *The Tempest:* "Our revels now are ended. These our actors, as I foretold you, were all spirits, and are melted into air, thin air; and like the baseless fabric of this vision . . . leave not a rack behind."

2. John Caldwell Calhoun—a former vice president of the United States—was a senator from South Carolina. Lewis Cass, a former brigadier general in the U.S. Army and governor of the Michigan Territory, was a senator from Michigan and in 1848 an unsuccessful presidential candidate. John Adams Dix, a senator from New York, was to become secretary of the Treasury during the Buchanan administration and a general in the U.S. Army during the Civil War.

George B. McClellan to Joseph G. Totten

WEST POINT

1 MARCH 1849

One case of varioloid[1] appeared in the Engineer Company four days ago. Eight or ten cases of the small pox and three of varioloid have occurred amongst the band and other enlisted men besides others of each order as in their families, during the last month.

These diseases are still prevailing in the vicinity of the Barracks of the Engineer Company, and it is not possible to move the Company into other quarters in order to avoid the influence of the contagion. There are no accommodations whatever for the sick of the Company, should the disease spread. I have applied to the Superintendent for some detached building to serve as a post hospital for my command, and it is not in his power to provide one, so that, in the event of this disease spreading, I shall absolutely be obliged to keep those afflicted with it in a room heretofore occupied as quarters directly in front of the Barracks which would be an act of cruelty towards the well men that I cannot voluntarily commit while in command of the Company.

I have the authority of Doctor Cuyler[2] for stating that, if the disease again breaks out amongst the men, they must suffer very much from the want of proper accommodations.

Even in cases of ordinary sickness, such as fevers and c., our men are entirely without proper hospital quarters. The room used as a post hospital is about twelve feet by fifteen feet, and the men, when sick, generally have to remain in their quarters, already too crowded for well men. Such being the state of affairs, and it being certain that neither our own Barracks nor a Hospital for enlisted men can be completed until some time during the year 1850 at soonest, I have the honor to submit to the Department the proposition that the Company be ordered to Fort Schuyler, at least until quarters and hospital are ready at this Post.[3]

I yesterday received a letter from Captain Cullum concurring fully in my opinion in this matter, and stating that Captain Benham[4] informed him that the men could be comfortably quartered at Fort Schuyler without any inconvenience to the Engineering operations about to be carried on there.

I may take this opportunity to observe that, independently of all considerations in relation to sickness, the Barracks are altogether insufficient for the company, and the men are altogether too crowded; and, if the Company should be filled to the number allowed by law, I do not hesitate to say that it will be absolutely impossible to accommodate them in these Barracks, yet the importance of filling the Company to carry on our Engineering drills to any advantage is self evident.

Men will always be discontented when they are very much crowded in their quarters, so that I regard it as of the first importance for the preservation of discipline, that ours should be less crowded than they are now. These men are expected to study and be taught. How is it possible for their officers to do justice either to themselves or their men when situated as we now are?

The casemates at Fort Schuyler would afford abundant room to the company and its officers—the ground there may be easily worked for sapping and mining purposes, which is by no means the case here, for there is no place at or near West Point where the ground is at all fit for such uses. The facilities for using the ponton train are much greater there than here.

I would beg leave to call the attention of the Department to the fact that the subalterns of this Company are quartered at a great distance from the Company and that we are subjected to great inconvenience in going so far to Barracks, for our duties call us there very often, and the absence of the officers from the vicinity exerts a bad influence on the discipline of the Company. I have no desire to urge an unreasonable or improper request, but I think that the facts I have stated render the proposition I have submitted both reasonable and necessary. At all events, whatever may be the decision in the matter, I shall have discharged my duty in proposing and urging what I regard as the best means of providing for the safety and health of the men under my command.

I am, sir, very respectfully yr. obdt. svt.,
Geo B. McClellan
Lt. of Engrs., Cmdg. Eng. Comp.

1. Varioloid is a mild form of smallpox.

2. John Meck Cuyler entered the U.S. Army on 1 April 1834 as an assistant surgeon. Following service in the war against Mexico, he was appointed as chief medical officer at the U.S. Military Academy, achieving the rank of major on 16 February 1847.

3. Fort Schuyler, positioned to protect New York City from naval attack, is located at Throgs Neck in the southwestern portion of the Bronx at a point where the East River meets Long Island Sound. Fort Totten faces it on the other side of the river.

4. Henry Washington Benham graduated first in the class of 1837 and was commissioned into the Corps of Engineers. During the war with Mexico, he was attached to Zachary Taylor's army. He would later serve under McClellan as a brigadier general of volunteers in the Army of the Potomac.

Frederic Augustus Smith[1] to George B. McClellan

WASHINGTON, D.C.

14 MARCH 1849

Dear McClellan,

The amendment to the army bill has passed, requiring the discharge on application of any noncommissioned officer, private, or musician of the Company of Sappers and Miners who may have served in the War with Mexico, but by one of their usual blunders it says "the company &c. raised under the Act of *May 9th 1848*" instead of *May 15th 1846*. As the intention, however, is plain, the General [Totten] has recommended to the Secretary of War to order the discharge from the *existing* company of the persons alluded to in the act.

The Secretary has not acted yet and may not for several days, so I thought I'd send you word in advance that there is a little or no doubt that he will give the order. You can go on and be making out the discharges. They will probably all apply, and if not, you should advise them to do it, for if they re-enlist the same day they get 3 months' extra pay and their bounty land.

I have been talking with McDowell,[2] too, about the non-commissioned officers. He says, according to the common if not the universal practice of the Army, the noncommissioned officers when [they] re-enlist the same day, continue their offices and their warrants will not be affected, so that an old Sergeant loses nothing of priority of warrant.

In the Pay Department, where you know the appointments are reviewed every four years, the officers rank according to their *original* appointment—a precise parallel case, and he tells me this is fixed.

I suppose you know that your Brevet and G. W. Smith's are changed, Smith's captaincy to date from Contreras and yours from Chapultepec.

Fraser and Barnard have both Brevets. Generals Jones, Towson, Gibson, Col. Talcott and Dr. Lawson *all* the heads of Bureaus here except Abert are brevetted. Mordecai and Baker of the Ordnance also. Brevet Captain Sprague has another Brevet. Turnbull and Hughes have another.[3] The fact is, the only trouble is to know who *isn't* brevetted. I am expecting myself, like Mr. Nickleby, every day, a hot express on horseback from the Senate to inform me that I'm a Colonel.[4] I shall wait a while, however, before I order my new uniform. The order will be out before long.

yours, etc.,
F. A. S.

The certificates of merit will doubtless be given to the Artificers recommended as well as the Privates, but as the list will probably not be made out for months, you had better tell those who go to leave with you a letter to the Adjutant General stating where the certificate is to be sent. These letters you and Cullum will certify to be genuine. The list of those recommended you have—Cullum has at all events.[5]

A law has also passed this session granting these certificates to the *non-commissioned officers,* and I believe, musicians. Our non-com officers will be cared for.

1. Frederic Augustus Smith graduated first in the class of 1829 and was commissioned into the Corps of Engineers. During the war with Mexico, he served as instructor of practical engineering at West Point and was in charge of recruiting and instructing replacements for the engineering company. Totten to Smith, 3 Sept. 1846, McClellan Papers, Library of Congress.

2. Irvin McDowell graduated from West Point with the class of 1838 and returned from 1841 to 1845 as an instructor of tactics. During the Mexican War he served as aide de camp to Brig. Gen. John E. Wool, earning a brevet to captain for his conduct during the Battle of Buena Vista. He is best known for his stinging defeat at the hands of the Rebel army under P. G. T. Beauregard at Manassas, Virginia, 21 July 1861.

3. Col. John James Abert was the commanding officer of the Corps of Topographical Engineers.

William Davidson Fraser graduated first in the class of 1834 and was commissioned into the Corps of Engineers. He was brevetted to major for "gallant and meretricious conduct while serving in the enemy's country." John Gross Barnard graduated second in the class of 1833 and was commissioned onto the engineers. He was brevetted to major for his conduct during the Mexican War.

Roger Jones, adjutant general; Nathan Towson, paymaster general; and George Gibson, commissary general of subsistence, were brevetted to major general "for meritorious conduct particularly in performance of [their] duties in the prosecution of the war with Mexico." George Talcott, chief of ordnance, and Thomas Lawson, surgeon general, were brevetted to brigadier general. All of these brevets were issued on 30 May 1848.

Alfred Mordecai, who graduated at the top of the West Point class of 1823 and commissioned in the Corps of Engineers, was brevetted to major of ordnance on 30 May 1848. Rufus Lathrop Baker of the ordnance department was brevetted to lieutenant colonel on the same date. Charles Jeffries Sprague, who had attended the academy but did not graduate, was a first lieutenant in the Ninth Infantry when he received a brevet to captain on 20 August 1847 for gallantry at Contreras and Churubusco. No record of the other brevet of which Smith writes exists.

William Turnbull, class of 1819, was brevetted to colonel of topographical engineers for his performance during the battles around Mexico City. George Wurtz Hughes was brevetted to major for his performance during the Battle of Cerro Gordo and to lieutenant colonel for meritorious conduct during the war.

4. In Charles Dickens's 1838–39 novel *Nicholas Nickleby,* the hero is unrealistically optimistic of better fortune.

5. Certificates of merit entitled their recipients to an additional two dollars in pay per month, and G. W. Smith had recommended twenty-seven of the enlisted men of the company of engineers for such certificates in recognition of their conduct in the battles around Mexico City. On 21 December 1847 General Totten forwarded his recommendations to the War Department, but by the end of 1848 had yet to win approval. "A large portion of the certificates have, up to this time, been withheld," he complained to Secretary of War Marcy, "from a misapprehension which, I trust, will no longer be suffered to work such wrong—weighing down the spirits of these fine soldiers under a feeling of neglect and injustice." John L. Smith to Winfield Scott, 27 Aug. 1847, in *Report of the Secretary of War*, 30th Cong., 1st sess., 1847, H. Exec. Doc. 8, 354, 429–30; *Report of the Secretary of War*, 30th Cong., 1st sess., 1847, H. Exec. Doc. 1, 281.

George B. McClellan to John H. B. McClellan

WEST POINT

22 APRIL 1849

My Dear John,

How come on things in Philadelphia? I hope you are not as stationary as we are here. Nothing new has turned up here for a long while, and we are patiently awaiting some new chance for a respectable grumble.

Old Zack bids fair to furnish the next subject. It seems probable that he will not order General Scott to Washington, still keeping up the old grudge against him for taking away the Regulars before Buena Vista. To be sure, it is a very high compliment to the Regulars, but it is a small, very small, thing in him to allow personal pique [to] deny us the right of having our proper Commander in Chief in his position. As things are now, the Army is virtually commanded by the Adjutant General, a miserable old fogie—so great an old fool that the Army would feel infinitely [obliged] to Father Abraham, or Old Scratch, if he would take him to his bosom in the shortest possible time.

So Charles Albert is well whipped. Old Radetzky with his Austrian discipline has been too much for the chivalrous Prince with his raw, cowardly Italians. Charles Albert deserved better fortune, for altho' not a fine General, he has shown himself a brave man.[1]

John Bull, having, of course, soundly thrashed the Seikhs,[2] is now fully at leisure to do so much for us, if we should choose to amuse ourselves by interfering in Canada affairs.[3] If Brother Jonathan feels disposed to have his skull cracked, I hope truly that his invincible volunteers may have the honor and pleasure (?) of opening the ball. I do not think that more than one campaign, conducted by Volunteer Generals and *citizen soldiery*, would be necessary to convince even the sovereign people that the "mercenaries" are good for something beyond eating Government rations.

I wish they would send that sword and be done with it. I don't want to come to Philadelphia until it is over. I dread such scenes beyond description and would much rather go into another battle than be concerned in any such affair. I wish sincerely that they had never thought of it. If you can do it quietly, just have it over at once—and remember *no newspaper fuss* "as you love me."

I went down to New York Friday week to meet G. W. He looks very well, indeed. Stuart is still here. Capt. Cullum has not yet returned, and I am still in command of the Company.

Tell Arthur that I received his letter and will answer it tomorrow or next day. Give my love to Mother and Maria and Mary. Do write to me soon and believe me ever,

your affectionate brother,

Geo B. McC

1. Charles Albert, the duke of Savoy, Piedmont, and Aosta and king of Sardinia, was a leader in the cause of Italian unification, fighting against Austrian power in Italy in the First War of Independence (1848–49). The Austrian army under Field Marshal Count Josef Radetzky von Radetz decisively defeated the Italians at the Battle of Novara, 22–23 March 1849, forcing Charles Albert to abdicate in flavor of his son Victor Emmanuel II.

2. The Second Sikh War broke out in 1848, precipitated by the murder of two British officials in Multan. In retaliation the army of the East India Company invaded the Punjab, winning a series of decisive engagements that resulted in the annexation of the Punjab into the company's domain.

3. Canadian elections in January 1848 chose a reform candidate as premier. Tory opposition resulted in massive rioting, however, and on 25 April 1849, only two days after McClellan wrote this letter, a mob burned the Parliament buildings in Montreal. Amid this furor many influential Americans favored annexation of Canada to the United States. Winfield Scott, for example, predicted on 29 June 1849 that two-thirds of the American people would rejoice at the admission of Canada, "and the other third soon would perceive its benefits." The uprising was suppressed, however, and no overt interference from American citizens took place. Kenneth R. Stevens, *Border Diplomacy: The* Caroline *and McLeod Affairs in Anglo-American-Canadian Relations, 1837–1842* (Tuscaloosa: University of Alabama Press, 1989); Scott to John George Hamilton, in Scott, *Memoirs,* 2:601–603.

George B. McClellan to Isaac I. Stevens

WEST POINT

APRIL 1849

The detachment of laborers (artillery) stationed here are to be transferred to the Engineer Company,—at least so many as may be necessary to fill up the Company. On our Company, then will it devolve to do all the police of the Point, to make the roads, drive the carts, feed the oxen, work in the blacksmith and carpenter shops, etc., etc.—in plain terms, the engineer company is destroyed; it has become a company of mud-diggers; it will no longer be an Engineer Company, for it will be impossible to do military duty, and no instruction in the duties of engineer troops can be given them. The object of the whole business is to get [Bvt. Maj. William H.] Shover's Company of [the Third U.S.] Light Artillery ordered on here, and we are sacrificed to attain that object.

It is a matter that concerns equally all the officers of our corps. We are disgraced if this order is allowed to remain in force, and I beg of you to use whatever influence you may possess in Washington to have the order rescinded, and the company ordered away from here.[1]

1. Stevens, *Life of Isaac Stevens,* 238.

George B. McClellan to Joseph G. Totten

WEST POINT

1 MAY 1849

Sir,

I have the honor to submit to the consideration of the Department the following memoir in reference to the Engineer Company. I have served with the company from its organization to the present time. I was its 2nd Lt., and actively engaged in the organization at West Point before leaving for the field. I served with it during the whole time that it was in Mexico and have been constantly on duty with it since its return. I can therefore fully say that I know what skill and knowledge Engineer Troops should possess, and on what points they should be instructed in order to fit them for their duties in time of war. In addition, I have, after a long experience, become fully alive to the unfitness of this Post as the station of the Company and deem it my duty to make the Chief Engineer aware of the true state of the case.

There are two distinct objects to be fulfilled by the Engineer Troops of our service. 1st To furnish to our armies in the field a corps of officers and men well skilled in both the theory and practice of the operations of sieges, the character of field intrenchments, military bridges, and the other important duties of such troops. 2nd To aid, or rather provide the means of diffusing, a certain amount of this knowledge amongst the other Corps of the Army.

I propose to examine whether either or both of these objects can be best, or even well, fulfilled at West Point. The importance, rather necessity, of their fulfillment is self-evident to every military man.

With regard to the 1st: In order to accomplish this, several things are necessary. The quarters should be comfortable and the officers be near the men, that the proper discipline may be maintained and the men be contented, they should be at a post by themselves, or, if that cannot be, at a place where they are the most *important* men and can obtain whatever may be necessary without trouble, or clashing with others. The work of the sapper and miner should be carried on upon the glacis of a permanent work in order that the relations between the attack and defense may be clearly seen.

But little benefit can ever arise from running saps, constructing batteries, cavaliers, &c., or running mines in a field where there is nothing to indicate the object or direction of these different works. The soil should be of the most favorable character, that is, totally of such a light nature that the labor expended may be the least possible, and the works when completed present the handsomest appearance.

As to the *Ponton-Train,* it should be thrown *over* a stream, or arm, of moderate width, say from 100 to 200 yards, with a depth of water at least ten feet, with a moderate current, in order to stretch the cables well. Good banks, affording proper facilities for placing the abutments and arranging the requisite depots of material used in the construction.

At the Point there are now no proper quarters for our Company, when filled to the standard of 100 men, which number is necessary to carry on our drills to advantage. Our present Barracks cannot properly accommodate more than fifty men. There are no conveniences for sanitation rooms &c. The officers are far removed from the Company. The proposition of occupying one of the Cadet Barracks is unadmissable for many reasons. There are no conveniences of kitchens, washing arrangements, privies, &c. The men would be brought in close contact with the Cadets, which would be an insult to the latter and would be in the highest degree prejudicial to our own men. The men would necessarily be deprived of the liberty and freedom of motion around the barracks, and could not indulge in any of the amusements which should be so much encouraged, that serve to pass away the time and make them contented, because they would interfere with the studies and amusements of the Cadets. Besides, it would justly be received with indignation by the officers of the whole army and would make our troops and Corps unnecessarily disliked.

Our new barracks, it is evident, cannot be finished, at soonest, before the winter of *1851*—probably the summer of 1852. From the fact of there being no Quarter Master here with Quarter Master's funds, we are often put to slight inconvenience that might otherwise be avoided.

There is no Permanent work nor glacis here and the ground is everywhere of such a rocky nature that it is literally impossible to sink shafts or run galleries, and the works of the sap, batteries, &c. can only be carried on at the expense of an amount of labor entirely too severe to produce good results.

The river is not fitted for the drill with the Ponton Train; it is either too shallow or too deep. If we used a small number of pontons, say fifteen or twenty, it is necessary to run them out towards the middle of the stream, thus making not a bridge but a floating wharf, and without any current to stretch the cables. If we use about thirty-five pontons, we can run from a small dock or front of our Barracks to a *rock* headland, about 220 yards from it, the water varying, by soundings, from four feet to six feet and with *no current* and altogether but a *pis-aller,* entirely unfit for our purposes.

Fort Schuyler unites all the advantages that can be desired, a good glacis, a fine work, ample quarters in the casemates for men and material, until Barracks and sheds can be built, a *separate* post, within reach of New York whence supplies and means of all kinds can be derived, a central and accessible position, whence the company may, in case of war, be at once moved to any part of our frontiers (and, lastly, it is a sufficiently near to West Point to allow us to aid in the instruction of Cadets.)

With regard to the second object, it may be attained in three ways. 1st, by allowing Cadets to see the inconsiderable and imperfect work that we can do at the Academy.

2nd, By being stationed at a proper post where all the works of an attack, and their object, can be seen by the graduating class, ordered there for the purpose, and which officers in general can [illegible] with advantage.

3rd, By a school of practice in connection with other arms of the service, to which the graduating class may be ordered, where many officers will be stationed, and which all may visit.

The 1st cramps the efficiency of the Company and could benefit the Cadets but little, for local reasons already given and which may not be avoided. We could not do enough to make it an object for officers to visit our works, and our efforts could be cursed with the ultimate decline and failure that are the results of all things commenced on a petty scale. Without the possibility of improvement, and the Company and its efficiency could easily degenerate into nothing.

The 2nd is apparently the most feasible and best plan at present, and would probably be the germ of the 3rd. Were we favorably situated the pride and interest of both officers and men would be excited, for we could exert ourselves to advantage, and I do not doubt but that the results of our labor would be a system of instruction and set of works and a Corps of Engineer troops as good as those of any service in the world.

I should propose that the graduating class be ordered to our post from the 1st of July to the 1st of September, as officers, or at least with fewer restrictions imposed upon then than when Cadets, under the command of the officers of the Company, and not to receive their diplomas and appointments until thy had passed satisfactorily through the course of practical engineering. They could then have the three months' furlough in the fall and join their Regiments on the 1st of December, instead of the 1st October, with a good knowledge of the character of Engineer Troops.

The 3rd plan would be altogether the best, could it be at once adopted. It would infuse new spirit and efficiency into the whole army and would tend to destroy all ill feeling between different Corps, by bringing them into contact. I think that the immediate adoption of the 2nd plan would furnish a most admirable basis of a general school of practice for all arms of the service.[1]

1. Totten's response to McClellan's "memoir" is not known, but in his 1850 report to Secretary of War Charles M. Conrad, the chief engineer observed that "the company has been assiduously employed during the year in acquiring the instruction, both theoretical and practical, necessary to the proper performances of their duties, and, while thus employed, has afforded the means of giving to the cadets most important aid in practical field engineering." He expressed his "extreme satisfaction" at the execution of the company's duties at West Point and remarked favorably on "the thorough soldier-like deportment of officers, non-commissioned officers, and men." *Documents Accompanying the Report of the Secretary of War to the President,* 31st Cong., 2d sess., 1850, H. Exec. Doc. 1, 269.

Isaac I. Stevens to Joseph G. Totten
1849

By law, the engineer company is restricted to one hundred men, a number entirely inadequate even to the duties of peace. . . . The remedy I would propose is this: Let the utmost care be exercised in enlisting men. Let no man be enlisted who cannot in due course of time be made a non-commissioned officer. Let there be in no case transfers from other branches of the service. Let the whole strength of the officers of the company be applied to discipline and instruct the men, so that in time of need we shall have a band of splendid non-commissioned officers, the peers of [Thornley S.] Everett and [David H.] Hastings and [Samuel Henry] Starr,—men who have received commissions for their gallant services in Mexico, and each of whom, had Smith and McClellan and Foster fallen, could have gloriously led on the company to its duty.[1]

1. Stevens, *Life of Isaac Stevens,* 238.

George B. McClellan to John Hill Brinton McClellan

WEST POINT

UNDATED

I received yesterday, my dear John, your kind letter of Monday. As I had no idea of turning *Banker,* I have no regret, personally, that the free banking bill did not pass. I supposed that it was a respectable business, like the banking houses of Europe.

Be assured that I've no idea of jumping headlong into anything. All that I have made up my mind to is this—that I will do my utmost to find some business which will enable me to leave a profession which has nothing to induce me to remain in it. I would not for the world have you suppose for a moment that I do not like my profession in the abstract. God knows I do. I have followed it since fifteen, and it is all I know in this world. In fact, all I care for, but under our Government nothing is to be hoped for in it.

Our Government may be a very fine one for civilians; it was not intended for military men. The more truly one likes his profession, the more soldierly pride he has, the more disgusted is he with our service. Could I sink down at once with indifference, stolidity, and foggyism, I would be perfectly content with the Army. As it is, it is a continual heaping of coals on one's head. I am almost afraid to read history or anything appertaining to my profession because the more I know, the more are my eyes opened to our own wretched condition.

We are entirely dependent upon a Congress whose guide is party politics, not justice or the good of the country. We can never hope to have a proper army organization except in the event of a civil war. Heaven keep me from staying in the service long enough to long for that and disunion as a chance, or [illegible] of ameliorating our condition.

You speak of a professorship. The only case in which I would accept a professorship would be if it were such that it would give me a respectable support and at the same time enable me to study Law. I would take such an one, tho' I think it would be hard to get. But as to accepting one with the intention of making it my mode of life, I would never dream of it. The Commanches may scalp me first.

I believe I can tell you in a few words all I wish and feel. My desire is to get into a business which will enable me, as many men see and hear of, to make an independence in a reasonable time, so that I may travel and do what I please. My ambition is to study law, not for the law itself, or the desire to follow the steps of the thousand and one pettifogging lawyers in starvation now, but because I know that it is in this county the surest road to eminence.

I do not see how it is possible for me to support myself while studying law. Therefore I give it up. If the other does also fail, after an honest and fair search, I can then remain comparatively [illegible], and shall form all my efforts to filling such a place in the opinions of my comrades—that is, in desiring to do so, for I don't care much for anybody's opinions, as long as I am in the right—as to give me a better chance in the coming war. A desperate, desperate hope! If I stay in the Army, I shall soon he a worthless growling foggie, despised by myself, and abused by all under me. Such appears to be the ordinary progress of affairs with us.

I've grumbled long enough to make a whole nation feel disgusted, so I'll wind up by saying that I've not yet heard from Saunders.[1] I will do my best to be ordered with him. Failing in that, I will apply for a leave for a few months, spend it in Philadelphia trying to form up something. That over, I shall go to Texas with my good old General Smith. There's my programme. *Qu'en pensez vous*?

1. John Saunders graduated second in the class of 1834 and was assigned to the Corps of Engineers. He won a major's brevet at Monterrey and in 1848 was appointed as chief engineer for the construction of Fort Delaware on Pea Patch Island. McClellan was assigned to duty under him there in June 1851.

Final Diary Entry

WEST POINT

22 SEPTEMBER 1849

Mc. thinks that he's booked for an infernally monotonous life for the remainder of his natural existence and wishes he were back again in No. 2 Calle San Francisco.

AFTERWORD

For the three years following his return from Mexico, McClellan served as an instructor at West Point. In June 1851 he was transferred to assist in the construction of Fort Delaware on Pea Patch Island in the Delaware River, but less than a year later, on 5 March 1852, he was appointed engineer, commissary, quartermaster, and second in command of Capt. Randolph B. Marcy's exploring expedition. Fulfilling McClellan's desire to see "some little of this big world of ours," Marcy's expedition discovered the headwaters of the Red River and mapped what was then the largest unexplored tract of Texas.

Back in Arkansas on 28 July, McClellan received orders to report to Brig. Gen. Persifor F. Smith, commander of the Military District of Texas. As Smith's chief of engineers, McClellan accompanied the general on tours of inspection of frontier forts, oversaw a survey of the state's rivers and harbors, and at the invitation of his old friend Isaac I. Stephens, then territorial governor, directed a surveying expedition for a proposed railroad through Washington Territory to the Pacific Ocean.

McClellan was promoted to first lieutenant on 1 July 1853 and to captain, with reassignment to the First Cavalry, on 3 March 1855. That same month Secretary of War Jefferson Davis sent him to Russia to observe French and English military operations in the Crimean War. When he returned from Europe, McClellan designed the famous cavalry saddle that still bears his name.

Despite having achieved his desire to "start for the other side of the water," McClellan resigned from the army on 16 January 1857 to become chief engineer and later vice president of the Illinois Central Railroad; in 1860 he became president of the Ohio and Mississippi Railroad. During this period as well, on 22 May 1860, McClellan married Randolph B. Marcy's daughter, Mary Ellen.

On 23 April 1861, with the outbreak of the Civil War, McClellan wrote to his bride, "My previous life seems to have been dedicated to this great end."[1] The governor of Ohio appointed him a major general of volunteers. His victories at Rich Mountain and Corrick's Ford, now in West Virginia, won him national attention, and on 27 July 1861 Pres. Abraham Lincoln appointed him commander of the principal Union army in the East, which McClellan reorganized and named the Army of the Potomac. Despite having pledged in 1847 that he would "never say another word against General Scott," he carried on a vindictive campaign against his old commander, superseding him on 1 November 1861 as general in chief of the U.S. Army.

McClellan proved himself to be a superb administrator, organizing, equipping, and training his command to near perfection. Moreover he restored the army's morale after its humiliation at First Manassas. No other commander ever won the hearts of the officers and men of the Army of the Potomac as did McClellan. But as a commander he left a good deal to be desired, grossly overestimating the strength of the enemy, quarreling with the administration, and refusing to take the offensive against the inferior Rebel army at Manassas Junction. In Lincoln's memorable words, McClellan had "the slows."

Not the least of the reasons for McClellan's want of aggressiveness was his assessment of the commanders whom he now faced. Beauregard, Johnston, and G. W. Smith fought for the Confederacy at Manassas. "I know full well the capacity of the Generals opposed to me," he told Samuel L. M. Barlow, "for by singular chance they were once my most intimate friends." Only after receiving a direct order from the president did he launch his amphibious invasion of Virginia in March 1862, landing his 118,000-man army at Fort Monroe on the tip of the peninsula formed by the James and York rivers. His march up the Peninsula toward Richmond was repeatedly checked by much smaller Confederate forces under Maj. Gen. John B. Magruder and his old comrade in arms, General

Johnston. Following Johnston's wounding during the Battle of Seven Pines, 31 May 1862, command of the Confederate army briefly devolved upon G. W. Smith and then, on 1 June, upon Robert E. Lee. Despite, or perhaps because of, their close association in Mexico, Lee and McClellan shared a mutual disdain for one another. At the beginning of the Peninsula Campaign, McClellan had written to the president, "I prefer Lee to Johnston—the former is *too* cautious & weak under grave responsibility—personally brave & energetic to a fault, he is yet wanting in moral firmness when pressed by heavy responsibility & is likely to be timid & irresolute in action." One week later he shared with Secretary of War Edwin M. Stanton his assessment that "Lee will never venture upon a bold movement on a large scale."[2] As Stephen W. Sears has noted, "McClellan did not elaborate on how he had arrived at this singular appraisal; mercifully for him, it was never made public during his lifetime."[3]

Lee was equally blunt in his evaluation of McClellan. When early in his 1862 Maryland Campaign Lee announced to Maj. Gen. John G. Walker his intention of destroying the Baltimore and Ohio Railroad bridge at Harrisburg, Pennsylvania, and then turning on Washington, Baltimore, or Philadelphia, Walker was astonished. "Are you acquainted with General McClellan," Lee asked, by way of explanation. "He is an able general, but a very cautious one. His army is in a very demoralized and chaotic condition, and will not be prepared for offensive operations—or he will not think it so—for three or four weeks."[4]

In a series of battles around the Confederate capital in late June, McClellan was repulsed by Lee's bold counteroffensive, and the demoralized general withdrew his army to its transports and sailed for Washington. Lincoln, disgusted with McClellan's failures, relieved him of his duties as general in chief in July 1862 and transferred most of his divisions to Maj. Gen. John Pope's Army of Virginia. But when Lee administered Pope a resounding defeat at the Second Battle of Manassas, McClellan was restored to command of his army.

He fought a drawn battle with Lee's substantially weaker Army of Northern Virginia at Sharpsburg, Maryland, on 17 September 1862 but failed to exploit his strategic advantage. Lincoln therefore removed him from command for a second and final time. Lee wrote to his wife: "I hate to see McClellan go. He and I had grown to understand each other so well."[5] Yet quite remarkably, when asked in 1870 which of the Union commanders whom he had faced he considered the greatest, Lee emphatically answered, "McClellan by all odds."[6]

Left without a command or a suitable administrative position, McClellan entered politics and in 1864 ran as the Democratic candidate for president of the United States on a peace platform. He resigned from the army on Election Day, 8 November, and after his narrow defeat returned to civil engineering until 1878, when he was elected governor of New Jersey. McClellan died in Maywood, New Jersey, on 29 October 1885 and was buried in Riverview Cemetery, Trenton.

1. Stephen W. Sears, ed., *The Civil War Papers of George B. McClellan: Selected Correspondence, 1860–1865* (New York: Ticknor and Fields., 1989), 113.

2. Ibid., 248.

3. George B. McClellan to Samuel Barlow, 8 Nov. 1861, Barlow Papers, Huntington Library; Sears, *Civil War Papers of George B. McClellan*, 244–45; Stephen W. Sears, *To the Gates of Richmond: The Peninsula Campaign* (New York: Ticknor and Fields., 1992), 57.

4. Robert Underwood Johnson and Clarence Clough Buel, eds., *Battles and Leaders of the Civil War*, 4 vols. (New York: Century, 1887), 2:604.

5. Quoted in Stanley F. Horn, ed., *The Robert E. Lee Reader* (New York: Grosset and Dunlap, 1949), 263.

6. Robert E. Lee, *Recollections and Letters of General Robert E. Lee by His Son Captain Robert E. Lee* (New York: Doubleday, Page, 1904; reprint, with new introduction by Gary W. Gallagher, Wilmington, N.C.: Broadfoot, 1988), 416.

GLOSSARY OF MILITARY ENGINEERING TERMS

Quotations in the glossary are from George Elliot Voyle, *A Military Dictionary. Comprising Terms, Scientific and Otherwise, Connected with the Science of War* (London: W. Clowes & Sons, 1876), and from *The Oxford Essential Dictionary of the U.S. Military* (New York: Oxford University Press, 2001).

bastion: a salient fortification constructed at the angle of a defensive line, intended to provide flanking fire along that line.

berm: a narrow, level space at the outside foot of a parapet, intended to retain material that otherwise might fall from the slope into the ditch.

billhook: an entrenching tool used for clearing underbrush, shrubbery, and tree branches and for cutting material for gabions and fascines.

boyau: a ditch covered with a parapet, serving as a means of communication between two trenches, especially between the first and third parallels; also called a zigzag or an approach.

casemate: a chamber within the ramparts of a fortification to house a number of guns, embrasures being cut for them through the revetment.

cavalier: a work constructed in the interior of a bastion; its terreplein is elevated from eight to twelve feet above that of the rampart so that artillery there may command the surrounding ground within firing range.

embrasure: an opening cut in a parapet to enable defending artillery to fire through it.

epaulment: a parapet on the flank of a battery serving to protect the guns and the gunners; the mass of earth or other material that protects the guns in a battery both in front and on either flank.

fascine: a long, cylindrical bundle of brush, sticks, or rods firmly bound together and used in filling ditches or marshes, for the construction of batteries, or for strengthening the sides of trenches, embankments, or ditches.

gabion: a wickerwork basket filled with rocks, used as a temporary field fortification.

galleries: the excavations formed underground from the end of a shaft to enable the miners to reach the required positions for placing the charges; they are called "shafts" when vertical and "galleries" when horizontal.

glacis: a gently sloping bank extending from the forward face of a fort, exposing attackers to fire from within. The parapet of the covered way extends in a long slope to meet the natural surface of the ground so that every part of it may be swept by fire from the ramparts.

parapet: a protective wall or earthwork along the top of a trench. In fieldworks this is a mass of earth thrown up as a protection against an enemy's progress, the ditch from which the earth is excavated forming an additional impediment.

revetment: a retaining wall supporting a rampart. "The masonry support afforded to the banks of earth on each side of the ditch, backed interiorly by buttresses. The scarp and counterscarp are such revetments. In field fortifications, the materials used for the revetments are gabions, fascines, [or] sandbags."

sap: A covered trench made for the purpose of approaching a besieged place under the fire of the garrison. "A trench formed by sappers, [who protect] themselves by filling gabions and placing them as fast as possible along the intended line of the parapet."

sap-fork: A tool shaped like a boat hook sappers used to advance or withdraw a sap roller—two large, connected gambions—during construction of approach trenches.

terreplein: the top, platform, or horizontal surface of a rampart on which the cannon are placed.

traverse: an earthen mask similar to a parapet, thrown across the covered way of a permanent work to protect it from the effects of an enfilading fire.

SELECTED BIBLIOGRAPHY

Bauer, K. Jack. *The Mexican War, 1846–1848.* New York: Macmillan, 1974.

Beauregard, Pierre Gustave Toutant. *With Beauregard in Mexico: The Mexican War Reminiscences of P. G. T. Beauregard.* Edited by T. Harry Williams. Baton Rouge: Louisiana State University Press, 1956.

Cullum, George Washington. *Biographical Register of the Officers and Graduates of the U.S. Military Academy.* Vol. 3. 3d ed., Boston: Houghton Mifflin, 1891.

Eckenrode, H. J., and Bryan Conrad. *George B. McClellan: The Man Who Saved the Union.* Chapel Hill: University of North Carolina Press, 1941.

Frazier, Donald S., ed. *The United States and Mexico at War: Nineteenth-Century Expansion and Conflict.* New York: Macmillan, 1998.

Furber, George C. *The Twelve Months Volunteer; or Journal of a Private in the Tennessee Regiment of Cavalry.* Cincinnati: U. P. James, 1857.

Goetzmann, William H. *Army Exploration in the American West, 1803–1863.* New Haven: Yale University Press, 1959

Hassler, Warren W., Jr. *General George B. McClellan: Shield of the Union.* Baton Rouge: Louisiana State University Press, 1957.

Heitman, Francis B., comp. *Historical Register and Dictionary of the United States Army, from Its Organization, September 29, 1789, to March 2, 1903.* 2 vols. Washington, D.C.: U.S. Government Printing Office, 1903.

Hillard, George Stillman. *Life and Campaigns of George B. McClellan.* Philadelphia: J. B. Lippincott,1864.

Hitchcock, Ethan Allen. *Fifty Years in Camp and Field: Diary of Major-General Ethan Allen Hitchcock, U.S.A.* Edited by W[illiam] A[ugustus] Croffut. New York: G. P. Putnam's Sons, 1909.

Hudson, Leonne M. *The Odyssey of a Southerner: The Life and Times of Gustavus Woodson Smith.* Macon, Ga.: Mercer University Press, 1998.

Hughes, Nathaniel Cheairs, Jr., and Roy P. Stonesifer. *The Life and Wars of Gideon J. Pillow.* Chapel Hill: University of North Carolina Press, 1993.

Kenly, John R. *Memoirs of a Maryland Volunteer: War with Mexico in the Years 1846–7–8.* Philadelphia: J. B. Lippincott, 1873.

Keyes, Erasmus D. *Fifty Years Observation of Men and Events, Civil and Military.* New York: Charles Scribner's Sons, 1884.

Lee, Fitzhugh. *General Lee: A Biography of Robert E. Lee.* New York: D. Appleton, 1894. Reprint, with a new introduction by Gary W. Gallagher, New York: DaCapo, 1994.

Maury, Dabney H. *Recollections of a Virginian.* New York: Charles Scribner's Sons, 1894.

May, Robert E. *John A. Quitman: Old South Crusader.* Baton Rouge: Louisiana State University Press, 1985.

McClellan, George B. *McClellan's Own Story.* New York: Charles L. Webster, 1887.

Meade, George Gordon. *The Life and Letters of George Gordon Meade, Major-General United States Army.* 2 vols. New York: Charles Scribner's Sons, 1913.

Morrison, James L., Jr. *"The Best School in the World": West Point, the Pre–Civil War Years, 1833–1866.* Kent, Ohio: Kent State University Press, 1986.

Ripley, R. S. *The War with Mexico.* 2 vols. 1849. Reprint, New York: Burt Franklin, 1970.

Robinson, Fayette. *An Account of the Organization of the Army of the United States.* 2 vols. Philadelphia: E. H. Butler, 1848.

———. *Mexico and Her Military Chieftains.* Philadelphia: E. H. Butler, 1847. Reprint, Glorietta, N.Mex.: Rio Grande Press, 1970.

Rodenbough, Theophilus F. *Army of the United States: Historical Sketches of Staff and Line with Portraits of Generals-in-Chief.* New York: Maynard, Merrill, 1896.

Scott, Winfield. *Memoirs of Lieut.-General Winfield Scott, LL.D.* New York: Sheldon, 1864.

Sears, Stephen. *George B. McClellan: The Young Napoleon.* New York: Ticknor and Fields, 1988.

Smith, Gustavus Woodson. *Company "A," Corps of Engineers, U.S.A., 1846–1848, in the Mexican War.* Edited by Leonne M. Hudson. Kent, Ohio: Kent State University Press, 2001.

———. *General Patterson's Route of March: Letter from the Secretary of War, Transmitting a Report on the Route of General Patterson's Division from Matamoras to Victoria.* 31st Cong., 2d sess.,1850. H. Exec. Doc. 13, 2–9.

Smith, Justin H. *The War with Mexico.* 2 vols. New York: Macmillan, 1919.

Stevens, Hazard. *The Life of Isaac Ingalls Stevens.* 2 vols. Boston: Houghton Mifflin, 1900.

Stevens, Isaac Ingalls. *Campaigns of the Rio Grande and Mexico.* New York: D. Appleton, 1851.

Wallace, Edward S. *General William Jenkins Worth.* Dallas: Southern Methodist University Press, 1953.

Waugh, John C. *The Class of 1846: From West Point to Appomattox: Stonewall Jackson, George McClellan, and Their Brothers.* New York: Warner Books, 1994.

Wilcox, Cadmus Marcellus. *History of the Mexican War.* Washington, D.C.: Church News, 1892.

INDEX